Entropy and Exergy in Renewable Energy

Edited by Lin-Shu Wang, Wenping Cao and Shu-Bo Hu

Published in London, United Kingdom

IntechOpen

Supporting open minds since 2005

Entropy and Exergy in Renewable Energy
http://dx.doi.org/10.5772/intechopen.92472
Edited by Lin-Shu Wang, Wenping Cao and Shu-Bo Hu

Contributors
Tayeb Brahimi, Ion Paraschivoiu, Radian Belu, Nouha Bouchiba, Souhir Sallem, Mohamed Ben Ali Kammoun, Larbi Chrifi-Alaoui, Said Drid, Luis Fernando Garcia Rodriguez, Julian Ernesto Jaramillo Ibarras, Jorge Luis Chacon Velasco, Juan Diego Rosero Ariza, Andriele De Pra Carvalho, Sieglinde Kindl Da Cunha, Adrian Tuck, Hadi Nabipour Afrouzi, Yuhani Pamodha Wimalaratna, Jubaer Ahmed, San Chuin Liew, Chin-Leong Wooi, Md Bazlul Mobin Siddique, David Borge-Diez, Canberk Unal, Emin Acikkalp, Bimal Kumar Sarkar, Lin-Shu Wang, Peng Shi, Kamyar Mehranzamir

First published in London, United Kingdom, 2022 by IntechOpen
IntechOpen is the global imprint of INTECHOPEN LIMITED, registered in England and Wales, registration number: 11086078, 5 Princes Gate Court, London, SW7 2QJ, United Kingdom
Printed in Croatia

British Library Cataloguing-in-Publication Data
A catalogue record for this book is available from the British Library

Additional hard and PDF copies can be obtained from orders@intechopen.com

Entropy and Exergy in Renewable Energy
Edited by Lin-Shu Wang, Wenping Cao and Shu-Bo Hu
p. cm.
Print ISBN 978-1-83968-662-7
Online ISBN 978-1-83968-663-4
eBook (PDF) ISBN 978-1-83968-664-1

For EU product safety concerns:
IN TECH d.o.o., Prolaz Marije Krucifikse Kozulić 3, 51000 Rijeka, Croatia, info@intechopen.com or visit our website at intechopen.com.

Meet the editors

Dr. Wang is an associate professor emeritus of Mechanical Engineering at Stony Brook University. His research activities focus on thermodynamics, beneficial building electrification, and Gaia 2.0. His latest work are the 2019 book entitled *A Treatise of Heat and Energy* and the 2021 paper *Progress in Entropy Principle* (*Progress in Entropy Principle, as Disclosed by Nine Schools of Thermodynamics, and Its Ecological Implication* | IIETA). He also has several patents including intercooled gas turbine, turbocharged internal combustion engine, and low-grade heat management in buildings.

Professor Cao is a chair professor at Anhui University, China. He is a fellow of IET and a senior member of IEEE. He received a Royal Society Wolfson Research Merit Award in 2016, the Dragon's Den Competition Award from Queen's University Belfast in 2014, and the Innovator of the Year Award from Newcastle University, UK, in 2013. He has served as an editor for IEEE Transactions on Power Electronics, IEEE Transactions on Industry Application, IET Power Electronics, and several other international journals. He has published over 260 research articles and 12 edited books and book chapters. His research interests include fault analysis and condition monitoring of electrical machines and power electronics.

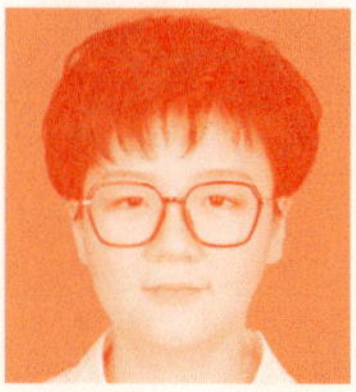

Dr. Hu is a postdoctoral researcher in the School of Control Science and Engineering, Dalian University of Technology. Her research interests include new and renewable power system dispatch optimization, battery energy storage, and big data analysis. Her latest work is the paper *Adaptive Time Division Power Dispatch Based on Numerical Characteristics of Net Loads* published in the *Journal of Energy*. She also has patents including power dispatch optimization, netload time-series analysis, and retired EV batteries.

Contents

Preface

Joule and Kelvin's introduction of the concept of energy in the middle of the nineteenth century sets the stage for the Second Industrial Revolution. The Revolution has been powered by fossil fuels, a form of stock energy that is rich in potential of entropy growth. The same richness in potential that makes it so useful, however, is the cause for high heat or CO_2 production in association with rich entropy growth, the unchecked accumulation of which, today, poses an existential threat to mankind on Earth. Cognizant of this fact, the worldwide communities have come to embrace the imperative for a twenty-first century transition from stock energy to renewable energy. This book aims to serve as a platform for scientists or engineers to disseminate their original research findings and their scholarship on development or literature review, as well as their perspectives on renewable energy.

A few preliminary words on energy and exergy: exergy, also known as available energy (one form of which is Gibbs free energy), is a concept derived from energy and entropy. Thermodynamically speaking, usefulness of energy can only be understood in terms of its exergy content: energies of low exergetic content are of little value. One cannot talk about energy without the language of exergy.

The theme of the book is that one cannot talk about energy's impact on our physical world without the language of both exergy and entropy, including the dual nature of the latter, and that understanding this dual nature is the key for order creation in the renewable energy era. The dual nature of entropy was demonstrated in a book (*A Treatise of Heat and Energy*, Dec. 7, 2019, Springer ISSN 0941-5122) in terms of the entropy principle as entropy growth selection principle and entropy-growth-potential (EGP) causal principle. This has bearing on one of the great scientific mysteries: how could life have evolved if the alleged tendency of the universe is to increase entropy until the universe heat death (Schrödinger's paradox)? A recent paper on progress in entropy principle (*Progress in Entropy Principle, as Disclosed by Nine Schools of Thermodynamics, and Its Ecological Implication* | IIETA) argues the dialectic case that whereas entropy principle as selection principle has been conventionally associated with inevitable increase of disorder, the principle as causal principle offers the explanatory framework for the emergence of orders of both technological kinds and biological kinds.

In addition to the audacious argument in favor of an entropy-order-emergence link, anecdotal evidence may also be found in Google Trends, as shown in **Figure 1**, for

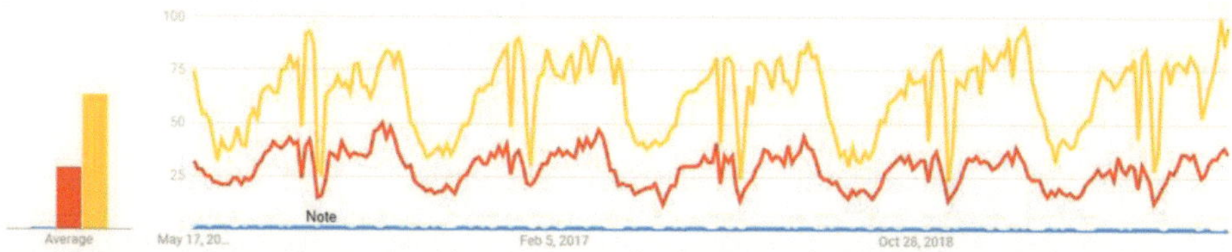

Figure 1.
Google Trends of renewable energy (in yellow), entropy (in red), and exergy (in blue) from May 17, 2015, to May 12, 2020 (accessed May 12, 2020).

natural affinity between inquisitiveness in entropy thinking, rather than in terms of exergy solely, and interest in renewable energy phenomena.

As architect James Wines wrote in 2008 Britannica Book of the Year, "The ultimate success of green architecture is likely to require that advocates achieve a broad-based philosophical accord and provide the same kind of persuasive catalyst for change that the Industrial Revolution offered in the 19th century." The same can be said about the transition from fossil fuel energy to renewable energy. The book and the author(s) of each book chapter are a small part of the collective "it takes a village" endeavor toward a post-industrial ecological revolution.

Lin-Shu Wang
Department of Mechanical Engineering,
Stony Brook University,
Stony Brook, New York, USA

Wenping Cao
Anhui University,
China

Shu-Bo Hu
Dalian University of Technology,
China

Section 1

General Considerations

Chapter 1

Systems-Thinking Framework for Renewables-Powered World

Lin-Shu Wang and Peng Shi

Abstract

Humans has experienced energy transitions throughout its history and the current transition from fossil energy to renewable energy is the latest example. But this latest example is different: rather than resulting from scarcity, this energy transition results from the threat of global warming—which is generally attributed to the short-term increasing of carbon dioxide in the atmosphere but also to the long-term heat threat posed by a warming Sun, according to the Gaia theory. Perspective appreciation of the nature of this combination necessitates for us to take a systems-thinking about the Earth system as a whole rather than the standard narrative of technical solution to our problem (of how to convert a small part of the abundant solar energy [including wind energy] into useful energy). Only by framing the energy transition as a part of dealing with the existential threat of global warming as *heat threat*, we are capturing the right perspective. Rather than any shortfall of energy—increasing carbon dioxide, heat threat, and collapse of Earth's ecosystems are the real threats. Cognizant of these is the beginning for humans to seize solutions to deal with the threats before it is too late.

Keywords: renewable energy, energy transition, global warming, heat threat from a warming Sun, systems thinking, entropy growth potential, carbon dioxide as a surrogate-indicator of the collapse of Earth's disequilibrium-ness, electrification of space heating

1. Introduction

This chapter is a perspective chapter—on the topic of renewable energy-resources or renewables, and how do we think about renewable energy and a renewables-powered world? One way to think about renewable energy is that it is a *form* of energy such as solar or wind that, unlike fossil energy that takes millions of years for its renewal, is being renewed diurnally. It should be emphasized that energy is never consumed (thus never needs to be renewed), and only the form of an energy, once it is transformed into heat form, needs to be renewed. Therefore, the vast difference between the times required for renewing the two kinds of energy is significant in talking about fossil energy and renewable energy. Historically, other kinds of transition from one kind of energy form to another kind of energy form have happen [1]. Nonetheless, the current transition is different; characterizing this transition from fossil energy regime to renewable energy regime as *energy* transition is, this essay argues, misleading and totally inadequate.

The total solar energy absorbed by Earth's atmosphere, oceans and land masses is approximately 3,850,000 exajoules (EJ) per year, of which it is estimated that the annual potential of solar energy converted into useful forms was 1,575–49,837 EJ still several times larger than the total world energy consumption, which was 559.8 EJ in 2012. If the annual need of energy by the world is about 560 EJ, only 0.015% of the total solar energy received by the Earth, the standard narrative of framing the energy problem is that we are blessed with the gift of 3,850,000 EJ/yr. energy input from the Sun and the challenge is how to convert a fraction of that energy input into useful forms of energy. Once that is achieved, problem solved!

I shall argue that this is a wrong way to frame the issue. The article's premise is that the idea of transition from fossil fuel energy to renewables is not just the transition from one form of energy resources to another form of energy resources; looking at renewables as a form of inexhaustible energies, or more correctly as a form of energies that are being readily renewed, is not sufficient reason for justifying energy transition: Instead,

the transition is necessary for overcoming an existential threat to humans rather than for solving the eventual shortfall of fossil energy (which cannot be readily renewed);

> *the transition is a transition from the notion of "energy and its consumption" to the notion of "entropy-growth-potential (EGP) management";*
> *that is, the transition is a transition from looking at our world in terms of "machines as machines powered by energy" to that in terms of "EGP-powered systems and their management."*

As President Kennedy said (see **Figure 1**), "In a crisis, be aware of the danger—but recognize the opportunity," the **purpose** of the essay is to make the point that this current energy transition is different from all previous ones [1] and this unique crisis of carbon-emission induced global warming is a crisis too "precious" to waste for us not to formulating solutions to deal with the human existential-risk [2] that is threatened not just by fossil fuels shortfall. EGP management and systems-thinking are two elements that'll help us to seize the opportunities.

To make the case, this essay makes use of a combination of **methods** including:

1. philosophical argument including argument against the machine-worldview and critiques on the scientific method or methodism are given in Section 2;

2. examples of buildings as homeostatic systems are presented in Section 3;

Figure 1.
J.F. Kennedy once said, "The Chinese use two brush strokes to write the word 'crisis'. One brush stroke stands for danger; the other for opportunity. In a crisis, be aware of the danger—but recognize the opportunity."

3. critical assessment of thermodynamics theory, which in its current form is deeply flawed; as well as

4. the example of the Earth system, the discussion is made in the context of the 1972 Gaia theory.

2. From the machine worldview to the systems worldview

Whether we are consciously aware of it or not, we have been conditioned in modernity since the 17th century by a foundationalist worldview [3–6], sometime called the Newtonian worldview of a determinist physicalist clockwork-universe. The full-pledged expression of foundationalism happened only in the nineteenth century. One timeline is the introduction of Laplace's Demon: "In the introduction to his 1814 *Essai philosophique sur les probabilités*, Pierre-Simon Laplace extended an idea of Gottfried Leibniz which became famous as Laplace's Demon," [7] which is a full expression of determinism. The second is one pointed out by Papineau [8], see also [3] that only with the advent of the law of conservation of energy which finally ruled out any force other than physical forces, the notion of a causally-closed physicalist universe became widely accepted. While a determinist universe solely in terms of physical forces was first suggested by Descartes and Leibniz in the seventeenth century, Papineau pointed out that later in the century Newton actually allowed possibility of non-physical forces [8]. The standard practice of calling the determinist physicalist clockwork-universe the Newtonian worldview, therefore, is actually inconsistent with the historical fact.

In any case, profligate acceleration of energy consumption also began in the nineteenth century. This made it possible to start the Second Industrial Revolution in which fuel-burning powered machines were the backbone of all industrial activities. So, it began a worldview in which a clockwork-universe dovetailed with viewing at such a universe made of all those fuel-burning powered machines: not only the whole universe is a machine, the universe is made of all those individual machines. For this reason, we shall call the "Newtonian" worldview a machine worldview.

In 1712, Thomas Newcomen invented atmospheric steam engine, the first practical fuel-burning engine which demonstrated that heat can be a source of power. The atmospheric engines were applied on site of coal mines, where the cost of coal was not an issue, for pumping water from mines. Their efficiency, i.e., the coals required for their operation, was not good enough for applications of atmospheric engines away from sites of mine. Those applications became possible when James Watt, in partnership with Matthew Boulton, made a critical improvement of atmospheric engines by separating the condensation process of steam from the main cylinder to another cylinder designated for condensing steam: instead of the main cylinder undergoing alternate heating and cooling (for the purpose of lowering the pressure in the cylinder thus the difference of atmospheric pressure and the resulting cylinder pressure is the force that produces power), the main cylinder is the heated cylinder while the separate condensing cylinder is the cooled one; whereby the lowering pressure in the main cylinder is obtained by opening the valve connecting the two cylinders. Thermodynamically speaking, the elimination of alternate heating and cooling reduces irreversibility, a key thermodynamic concept, in the operation. The great reduction in coal consumption made it possible for the Boulton-Watt atmospheric steam engines to be used as stationary powerplants away from coal mines. In 1776, Boulton said to Boswell, who was visiting him, "*I sell here, Sir, what all the world desires to have—power.*"

This historic technology advance initiated the First Machine Age with factories with power source not only free from the constraints of water and wind but also of magnitude unimaginably higher than animal, water and wind powers. For the first time in history power can be obtained reliably, independent of the capricious nature of water and wind. Power is where engine is, stationary ones or movable ones. Engine power augmented muscle power wherever engines and atmospheric engines are located—*separating* the power (that could drive factory machines) from the capricious nature. That was the beginning of the Second Industrial Revolution.

The Second Industrial Revolution would not be a complete revolution without another transformative technology, electricity. Most people give credit to Benjamin Franklin for discovering electricity. The invention of the electrochemical battery by Alessandro Volta in 1799 made possible the production of persistent electric currents. Hans Christian Orsted, and Andre-Marie Ampere separately, investigated electromagnetic interaction and described how electric currents through electromagnetic interaction could give rise to mechanical force and motion. It was Michael Faraday who discovered electromagnetic induction and demonstrated the phenomenon in the opposite direction, how motion through electromagnetic induction could give rise to electric currents. Thus, the production of power and motion could be used to generate electric currents, which could be transported over large distance with the invention of high voltage AC currents—separating the power from the engines. Central electricity powerplants now could power electric motors driving operations of factories, making possible for further *flexibility in siting* factories, which can be sited wherever within the reach of grid of a centralized powerplant.

All these machines, mechanical ones and electric ones, are fed with input of fuel-energy or electric energy and have specified output of work, or delivered heat energy, or delivered cold energy (i.e., heat removal), or value-added products. With defined input and output, and well-established relation between input and output, performance of the machines is defined in terms of efficiency; in the case of delivered heat-removal, "efficiency" is in the form of "coefficient of performance."

In the early 20th century, efficiency movement, a movement that sought to identify and eliminate waste, became the obsession of continuously improving operation in all areas of the economy and society [9]. The second law of thermodynamics and the concept of reversibility (and irreversibility as the cause of loss in efficiency) were the theoretical cornerstone of that movement. Augmenting human and animal muscle power [10] and improving in augmentation through continuous efficiency gain has been the reason for the singular transformation of the last three hundred years since the Enlightenment.

There is one category, however, that has so far escaped the reach of the Enlightenment and the success of science undergirded by the foundationalist worldview, the category of systems and complex systems. For example, a building is a complex system, the study of which has been greatly enhanced by computer simulation tool, such as DOE's *EnergyPlus*. But this new category is different in more fundamental way than just being more complex: they are systems instead of machines.

3. Buildings as examples of homeostatic thermal-systems

Astrophysicist Emden published in *Nature* [11] a short article in the form of puzzle or riddle:

> *Why do we have winter heating?*
> *The layman will answer: "To make the room warmer."*

> *The student of thermodynamics will perhaps so express it: "To import the lacking (inner, thermal) energy."*
> *If so, then the layman's answer is right, the scientist's is wrong...*

Emden correctly perceived no intrinsic relation between the "lacking energy" and making the room "warmer." Yet, the issue of energy for building applications is universally addressed in terms of energy efficiency. The truth is that, absent of an input-and-output relation, energy efficiency is meaningless.

ASME (American Society of Mechanical Engineers) noted in a 2013 study report,

Whereas the United States has made significant progress in increasing efficiency and reducing energy use in the transportation and industrial sectors of the economy, both building sector energy use and building system energy use have shown only modest reductions, well below what building owners and government policy leaders have hoped for. Automobiles, aircraft systems, and locomotion systems have all shown efficiency improvements twice that of building systems... (ASME Integrated / Sustainable Building Equipment and Systems (ISBES) Open Research Forum (ORF-1) April 24, 2013 Washington, DC).

Neither the movement based on efficiency improvement nor the green-building movement has produced the result that have been the intense pursuit of the first two decades of the 21st century. There are two possible interpretations of this 2013-report conclusion of lacking of progress in building sector: (1) there is some fundamental misunderstanding of what a building is and, as a result of that, we fail to find effective building solutions; (2) building energy efficiency is the wrong metric as progress indicator so that talking about lacking of efficiency improvement is a red herring (see below for an alternative performance-metric).

Both interpretations are correct. The best way to decode the Emden riddle begins with the recognition that a building is not a machine and it is not designed to have a product output. Instead, a building is a system, the "design goal" of which is in keeping the state of its existence within homeostatic ranges, in particulate within a temperature range. "The scientist" and "the student of thermodynamics" are wrong because they have been trained in viewing every system as machine instead of real system—and the performance metric of every machine is in terms of efficiency.

Architects appreciate partially the point in this way: building conditioning should not be based on machine-based solutions failing to see a building as a system as a whole. Addressing the building conditioning problem, "Albert, Righter and Tittmann" characterized the solutions of the three centuries this way as shown in **Figure 2**. ART depicted a 19th century building offering minimal thermal comfort. In the 20th century, the building conditioning was handled by machines of the First Machine Age resulting in, as ART depicted, a messy, incoherent set of devices. The point is that the machine-based solutions were conceived without a plan for maintaining a building as a system as a whole. This practice continues today. In the third depiction, ART suggested the building being maintained by renewable energies managed with mechanical assistance—basically it suggested an architecture-based solution that are known as green building solutions by USGBC. While the eventual success of transitioning from mechanical-engineering solutions to green-building solutions remains an open question, **Figure 2** correctly suggests that architectural societies and engineering (ASHRAE) societies need, in partnership, to look at buildings as systems, not machines.

Furthermore, the full implication of thinking in terms of systems must go beyond individual systems to think about both individual systems and how the individual systems, in the context of building systems, interact with each other and with "power-grid/powerplants-that-power-the-grid." Systems thinking is very much ecological thinking.

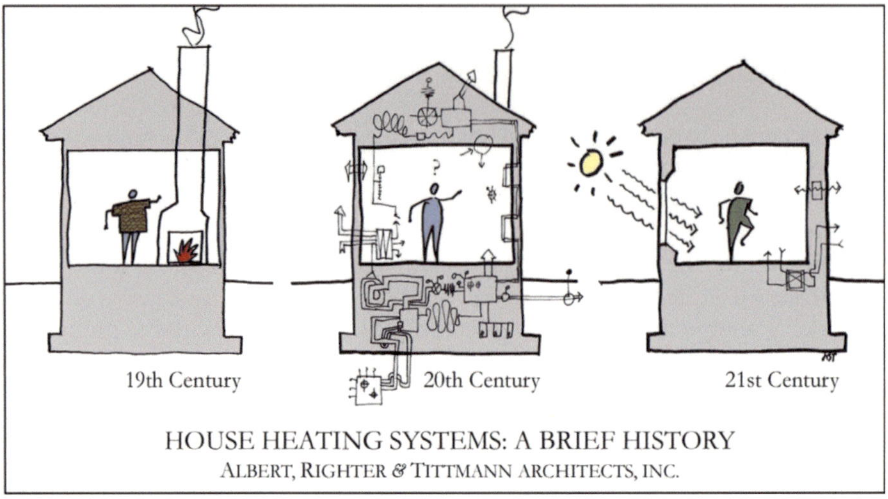

Figure 2.
Evolution of building heating systems over three centuries: From architectural solutions to mechanical-engineering solution to machine-assisted green solutions.

Electrical power is an energy carrier that can be powered by renewables. In recent years, wind power and solar electric power have become cost-wise competitive with traditional powers, and there is a consensus that electrification-of-everything is the best approach to achieve ultralow-carbon emission goal or zero-carbon emission goal—which is the ultimate objective of the current energy transition project.

Instead of being distracted with "increasing efficiency and reducing energy use in ... building sector," we have the perfect performance-metric for building, *carbon emission.* A recent study of the application of such metric unveils a very interesting finding.

A typical electric grid is powered by a mix of generators: baseload powerplants of nuclear, coal, and hydro; natural gas electric-generators; wind farms and solar farms; fossil-fuel peak-stations. The carbon emission related to electricity generation/consumption is strongly dependent on the actual mixture of the generators with various types of fuel sources. The study, a doctoral thesis [12], is based on the 2019 (hourly) time series data of electricity generation/consumption by source fuel type (NY ISO [13]) for determining hourly *carbon intensity*,

$$carbonemission = carbonintensity \times electricityconsumption \tag{1}$$

Because of high variation in the instantaneous mixture of generators, both hourly carbon emission, which depends on hourly fuel consumption, and hourly carbon intensity are highly variable.

Unlike the hourly generation/consumption by source data, the hourly fuel consumption data is not available from NY ISO. The only related fuel consumption data for electricity generation comes from US EIA electricity monthly database. From this database, the fuel consumption and resulting electricity generation data are available in monthly resolution in New York in 2019. Therefore, the efficiencies of various types of generators can be calculated. The monthly generation efficiencies from natural gas, petroleum liquid, and coal generators are up-sampled as constants and serve as denominator of the hourly electricity generation by source data to calculate the hourly consumption of the three fossil fuels. Then total carbon emissions from these fossil fuel consumptions can be calculated in hourly resolution with the EPA Greenhouse Gas Inventories.

Once the hourly carbon emissions data associated with electricity generation is obtained, the *time-varying carbon intensity* of electricity is calculated by dividing that with the hourly electricity generation/consumption data. One may wonder if the ultimate objective is to lower total carbon emission why one goes through the loop of dividing carbon emission data with electricity consumption data—for the purpose of multiplying the resulting carbon intensity with electricity consumption, again, to get carbon emission estimate. The answer is that carbon intensity is a function of existing grid based on current pattern of grid-wide electricity usage, whereas consideration of change in individual electricity consumption may be made for evaluating impact of such change on carbon emission. Such consideration may be made under the assumed carbon intensity, which will not change in short term. In short, with an assumed unchanging ***carbonintensity*** (e.g., in gray in **Figure 3**), estimate of ***carbonemission*** of different demand of electric usage (in yellow in **Figure 3**) can be made.

$$\text{estimateofcarbonemission} = \sum_{Year} carbonintensity \times changedelectricityconsumption \tag{2}$$

Calculated result of carbon intensity based on existing pattern of grid-wide electricity usage in the study [12] is reproduced here in **Figure 3**. Superimposed with carbon intensity in the figure is the simulated electricity consumption makeup of an individual building with air-conditioning as well as electrified space heating and domestic water heating shown in yellow (referred to as eHP). Note the high winter peaks of eHP as a result of winter space heating, whereas the standard common practice of a combination of air-conditioning and fossil fuel fired space heating and domestic hot-water heating has peaks, much lower ones, in summer only. So, when we multiply the carbon intensity with electricity consumption (electric demand), it is a very different electric demand makeup (from that determining the existing carbon intensity) resulting in very different carbon emission estimate.

The estimate of annual carbon emission is shown as **Figure 4**.

As a result of the peaks of carbon intensity (based on current grid with its operation outside the summer season being underused) and electricity demand of eHP are out of phase with each other, a 70% reduction in carbon emission (reduction from 7087 to 2214) is projected even with the current grid. Even with limited

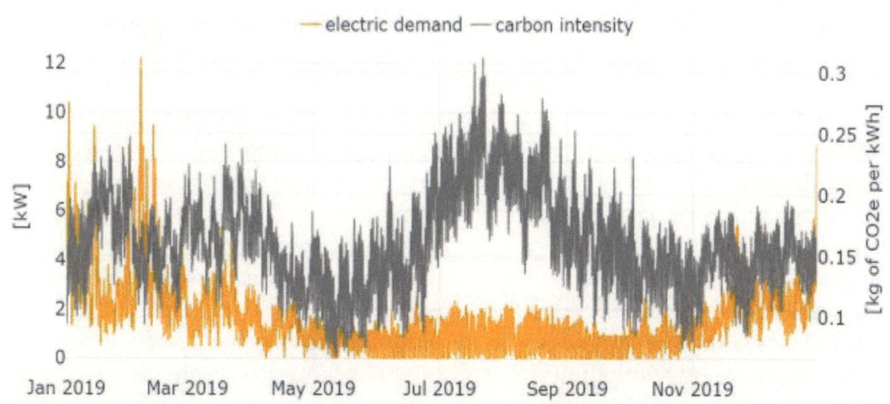

Figure 3.
Carbon intensity of existing pattern of grid-wide electricity usage and simulated electric demand of eHP (electrified space cooling & heating and DHW heating).

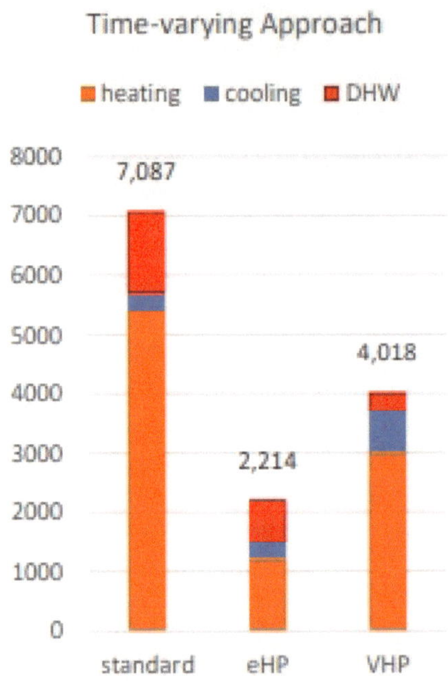

Figure 4.
Estimate of annual carbon emission of eHP in comparison with that of standard building conditioning.

penetration of renewables in our current grid, an instant drastic reduction in carbon emission can be achieved when each individual building goes to be fully electrified.

This is an example of the great potential of systems approach.

Another characteristic that differentiates systems approach from "Newtonian" machine approach is while the machine worldview is a static worldview, the systems worldview sees the world as a dynamic, changing, world. The finding of **Figures 3** and **4** make a compelling case for immediate deployment of electrification of space heating and domestic hot water heating—as6long as the extent of such deployment does not change "pattern of grid-wide electricity usage" in any significant way, i.e., the carbon intensity shown in **Figure 3** holds.

When market-deployments of electrification of heating reach significant market penetration, it'll change pattern of grid-wide electricity usage into one that manifests high winter peak demand. That is a problem which has to be solved—without which the progress of electrification of everything including heating will be halting to full stop resulting from a carbon intensity very different from the one shown in **Figure 3**. Both the cost advantage and the carbon emission advantage will vanish. Usage of energy storages both of electricity storages and thermal energy storage (TES) will be *crucial parts* of the solutions, investigation of the latter kind, TES, has been carried out in the study [12].

The building sector is a good example of systems approaches. Buildings, when they are considered as parts of grid system, are example of ecosystems. However, as ecosystems they are different from "ecological systems that are made of organisms" in a fundamental nature: unlike organisms that are active participants of the ecological systems having geophysical/geophysiological impacts on the non-living part of the systems (especially, see the Gaia discussion in Section 5), human inhabitants are passive components of building without defining building in a physical or ecological sense. While it has been suggested that standard theories of thermodynamics fail to treat the world in terms of systems thus fail to equip students of thermodynamics to have a solid grasp on the study of buildings, their shortcoming in dealing with ecological systems is much more serious. A case can be made that they totally miss the central question and its related core issues. A brief report on how this can be remediated is given in the next section.

4. The first law and the second law of thermodynamics

Theory of thermodynamics is constructed on the two laws of thermodynamics, the first law and the second law. Generally, students of thermodynamics consider that the first law is well understood but that the second law is the one that is difficult to comprehend. One obvious problem is that there are two versions of the second law: the original version formulated by Thomson (later, lord Kelvin) that of the universal degradation of mechanical energy [14], which'll be called, in short, *the energy principle*; the later version formulated by the Berlin School of thermodynamics by Clausius and Planck, which'll be called the entropy principle. The entropy principle has been universally acknowledged to be the true second law of thermodynamics.

There are multiple problems here. First is that though the entropy principle is accepted to be the true second law most students consider the meaning of the entropy principle to be encapsulated by the universal degradation of high-grade forms of energy, i.e., the energy principle. If the two principles are merely synonyms, that situation is acceptable. The problem worsens, therefore, because they are not. That demonstration can be made by showing that while the entropy principle is a universal principle, the energy principle is *not* a universal principle. Such a conclusion was reached by Planck [15]:

> *The real meaning of the second law has frequently been looked for in a "dissipation of energy." This view, proceeding, as it does, from the irreversible phenomena of conduction and radiation of heat, presents only one side of the question. There are irreversible processes in which the final and initial states show exactly the same form of energy, e.g., the diffusion of two perfect gases or further dilution of a dilute solution. Such processes are accompanied by no perceptible transference of heat, nor by external work, nor by any noticeable transformation of energy. They occur only for the reason that they lead to an appreciable increase of the entropy ([5], pp. 103–104).*

Details of Planck's argument has been worked out in a recent book, *A Treatise of Heat and Energy* [16], which concludes that mechanical energy degrades spontaneously not universally.

A good question would be why Planck's conclusion, which is of supreme importance, has not been more widely disseminated. *A Treatise* seeks to explain this situation by arguing that the entropy principle formulated by the Berlin School was a selection principle: the 19th century Berlin School was under the sway of the foundationalist mechanical-philosophy and the only kind of selection permitted by the entropy principle as a selection principle in accordance with the mechanical philosophy is selection based on physical necessity or efficient causation. As a result of the metaphysics of necessity as physical necessity alone, the inevitability of entropy growth infers the corollary of inevitable accumulation of heat, i.e., the energy principle.

In short, under the sway of the mechanical-philosophy, Clausius and Planck were not able to formulate a second law that Planck clearly realized, later on, should cast away the energy principle.

Before a new formulation of the second law for achieving that purpose is discussed, let us look at the first law: it turns out that we have misconception about the first law as well. Misconception about the two laws is intrinsically intersected: The first law was based on the mechanical equivalent of heat, i.e., the equivalence principle. The equivalence principle is the idea that all forms of energy during transformation from one form to another are conserved, i.e., all energy-forms are universally connected. But connection does not mean (necessarily) causation. That misstep was taken by Thomson and Clausius when they independently formulated the first and the second laws by giving causal power (the power that should be the purview of the second law) to the first law—making it too powerful depriving the second law of its rightful purview. The result of that misstep is a second law as a selection principle instead of a selection and causal principle.

This point was made in A Treatise. A more detailed discussion can be found in a new paper [17]. Back to the issue on the second law: a short account of a new formulation of the second law as it is related to the Carnot cycle is shown here.

The Carnot cycle can be interpreted differently from how it has been taken according to the conventional perspective: Instead of it as "an energy conversion of heat energy at T_A to mechanical energy," we consider the cycle as "the reversible event, in reference to a corresponding spontaneous event, of heat transfer from a T_A hot body to a T_B cold heat-reservoir." We thus begin, with the same setup enabling the Carnot cycle, by considering two "book-end" events,

1. The spontaneous event of the heat transfer process, the reference event, and

2. The reversible work production event of the Carnot cycle.

The book-end events define a *Poincare range* [15, 16]. Consider, first, entropy growth in the spontaneous event, in which the same amount of heat energy Q_A exiting the T_A heat body enters the T_B cold heat-reservoir. Therefore, the entropy growth in the universe is:

$$(\Delta_G S)_{univ} = \frac{-Q_A}{T_A} + \frac{Q_A}{T_B} \tag{3}$$

Whereas in the reversible event, a smaller amount of heat energy, Q_B, enters the T_B cold heat-reservoir resulting in zero entropy growth, in accordance with the entropy principle,

$$0 = \frac{-Q_A}{T_A} + \frac{Q_B}{T_B} \tag{4}$$

The heat energies exchanged with the T_B heat-reservoir during the two events are Q_A and $Q_A \bullet \frac{T_B}{T_A}$, respectively, as summarized in **Table 1**.

Rather than as a fraction of Q_A in accordance with the conventional perspective, note, in this new perspective, that the mechanical energy, (W_{rev}), is *preciously* the

Heat discharged to the T_B heat-reservoir	Event
	the spontaneous event
$Q_B = Q_A \bullet \frac{T_B}{T_A}$	the reversible event (According to a standard Carnot heat engine treatment)

Table 1.
Difference in heat discharged to heat-reservoir for the two book-end events.

difference between the amounts of heat energy added to the T_B heat-reservoir for the two events, given by:

$$W_{rev} = Q_A - Q_A \bullet \frac{T_B}{T_A} \tag{5}$$

Which is found to be the product of the universe's entropy growth in the spontaneous event according to (3), $(\Delta_G S)_{univ}$, and the temperature of the cold heat-reservoir, T_B,

$$W_{rev} = Q_A - Q_A \bullet \frac{T_B}{T_A} = T_B \bullet \left(\frac{-Q_A}{T_A} + \frac{Q_A}{T_B} \right) = T_B \bullet (\Delta_G S)_{univ} \tag{6}$$

Accordingly, we call the universe's entropy growth in the spontaneous event, $(\Delta_G S)_{univ}$, the "*entropy growth potential*" (EGP), $(\Delta_P S)_{univ}$ (for the reason articulated in the paragraph below):

$$(\Delta_G S)_{univ} = (\Delta_P S)_{univ} \tag{7}$$

Eq. (6), then, becomes,

$$W_{rev} = T_B \bullet (\Delta_P S)_{univ} \tag{8}$$

The logic of calling $(\Delta_G S)_{univ}$ "EGP" in (7) and (8) is that EGP is a *common property* (the term Poincare used in [16, 18]) of both events, as well as of all possible events in the Poincare range. This common property is the driver for enabling the *extraction* of a given amount of heat energy from the T_B heat-reservoir and converting it to mechanical energy of the same amount; in each case the amount for a specific event, though subject to the same "common property," is different; the maximum amount of extracted heat for a Poincare range is given by (8).

The same kind of demonstration on the idea of a common property has been made for systems in general, especially for isolated composite systems, in A Treatise and in References [17, 19]. We have, therefore, as a new part of the second law, *entropy growth potential principle*, in a general statement [19]:

> *for a given non-equilibrium system, the spontaneous event of the system approaching equilibrium state and the corresponding reversible event defined by the same initial and final states define its Poincare range; any event in the range shares the same common property of EGP, while the specific entropy growth is different dependent on the individual event in accordance with its individual causal necessity.*

Note that, for the existence of a system's EGP in association with change between the initial state and the final states of the system, physics does not require system energy change (though it often is associated with system energy change). Energy is NOT a necessary substrate for the existence of EGP. In contrast, physics (the second law) does require the system in its initial-state existence at non-equilibrium state. A good safe distance from equilibrium state is the defining condition for a system, any system, to be the driving force for making the world go around—the precise metric of which is its entropy growth potential, EGP.

In sum, the conventional formulation of the first law is too powerful depriving the second law, as a selection principle, of its rightful purview. We have reformulated the first law [17] by taking away the causal power of energy—and reformulated the second law [16, 19] as a selection principle (the inevitable growth of entropy) and causal principle (entropy growth potential principle).

5. Homeostasis in ecological systems, and the need to keep the Earth system cool

The conventional thermodynamics was formulated in the 19th century under the orthodoxy of the foundationalist mechanical-philosophy, in which the world is a machine made of machines. The theory is unable to deal with systems, especially ecological systems made of biological organisms.

Because the conventional theory is based on metaphysics of physical necessity, in dealing with complex systems with emergent orders a common theory of complex systems is known as maximum entropy production principle (MEPP). MEPP accounts for the emergence of local orders of individual complex systems by individual complex systems' ability to export entropy produced (grown) internally/locally to their surroundings. In the context of the Earth and ecological systems on Earth, such conclusion would predict an Earth ecosystem with greater and greater entropy corresponding with higher and higher global disorder.

A living organism becomes a dead organism by definition if its existence approaches a state of thermodynamic equilibrium or it exists in an environment that is approaching thermodynamic equilibrium. Because of that, a living organism as well as any complex system consisting of living organisms can only exist at states safely away from equilibrium. Aside from a metric-set of homeostatic ranges (for instance, temperature range), "far from equilibrium existence" (its metric is the entropy difference between entropy of the existing system and entropy of the system when it would approach thermodynamic equilibrium) is another defining characteristic of the *homeostatic state* of an organism. That is,

> ***homeostatic state*** = *metric-set of homeostatic ranges*, and *entropy difference of the system from system-at-reference-equilibrium-state.*

It is the latter defining characteristic that disqualifies MEPP, though it may be valid for complex physical systems such as climatic systems, for explaining the emergence of biological orders [19]. MEPP is a theory that is in full compliance with metaphysics of physical necessity. Ref. [17] puts forwards that by admitting causal necessity, the inference that the inevitability of entropy growth leads to the inevitable accumulation of heat, which is accepted and embraced by MEPP, is broken. Correspondingly, Ref. [19] puts forwards the thesis that emergence of biological and ecological orders requires admitting causal necessity as well. That is, the abandonment of mechanical-philosophy with its physical necessity stricture.

An example of this kind of consideration is the body of work on Gaia by James Lovelock, who applied far-from-equilibrium consideration to complex system consisting of living organisms. When he was a consultant at the Jet Propulsion Laboratory in Pasadena, CA, he was given the assignment of how to detect whether a planet harbors life. Lovelock began with the hypothesis that a planet as a complex system consisting of life—like a single organism—must be far from equilibrium or at radically disequilibrium state. Therefore, its atmospheric chemical composition must exhibit high concentrations of reactive gases, such as Earth's atmosphere which contains high concentration of oxygen and methane. Whereas, the static Martian atmosphere composing of almost entirely of non-reactive carbon dioxide is indicative of it being absent of life. Lovelock then took the next step by hypothesizing the "renewing" of these reactive gases to be a self-regulating mechanism of a planetary ecosystem. Lovelock together with microbiologist Lynn Margulis went further claiming the Earth to be in effect a superorganism, called Gaia (Lovelock, [20], Lovelock and Margulis, [21, 22]; Margulis and Lovelock, [23]). This version of Gaia, of a "living" complex system consisting of living organisms just like a single living organism, had received strong push back, when it was originally proposed, by biologists especially evolutionary biologists as unworkable in theory (Dawkins, [24]; Doolittle, [25]).

However, the idea of Gaia that all living things collectively define and maintain the conditions conducive for life through a filtering "selection" mechanism has since begun to receive acceptance [26] including Doolittle himself (see below). What is at issue is not the disequilibrium state of the Earth and that some kind of self-regulating mechanism for maintaining the state homeostatically (the former is a matter of physics and the latter is an observational fact of the Earth system), but how a "superorganism" acquires such a mechanism. Doolittle, in his reassessment of Gaia, put the matter this way (very different from his view of four decades earlier) as:

> *The Gaia hypothesis in a strong and frequently criticized form assumes that global homeostatic mechanisms have evolved by natural selection favoring the maintenance of conditions suitable for life. Traditional neoDarwinists hold this to be impossible in theory. But the hypothesis does make sense if one treats the clade that comprises the biological component of Gaia as an individual and allows differential persistence – as well as differential reproduction – to be an outcome of evolution by natural selection. Recent developments in theoretical and experimental evolutionary biology may justify both maneuvers [27].*

This new assessment on Gaia is a momentous step, which confirms the rejection of mechanical-philosophy—additionally, it makes the metaphysical presupposition that the world is made up of *natural kinds* such as atoms, molecules, and chemical elements, and *individuals* such as organisms, species. Clades, and Gaia. Whereas the former is characterized in terms of physical necessity, the latter in terms of physical necessity and causal necessity. The concept of natural selection was the revolutionary step taken by Darwin to finesse the teleological issue within the orthodoxy of mechanical-philosophy in biology. That was revolutionary and subversive. With the new momentous step, natural selection, which seemed to be a poster-boy of mechanical-philosophy, now undergoes its subversive transformation overthrowing the mechanical philosophy to include "survival of reproduction competitiveness" as well as "persistence as a result of global homeostatic mechanisms" [27].

One of Earth's homeostatic mechanisms is the mechanism to keep the Earth cool, according to Lovelock, in face of Sun's increasing solar radiative heat output. It is necessary to keep the Earth cool because:

> *It is vital for our survival that the sea is kept cool ... Whenever the surface temperature of the ocean rises above 15°C, the ocean becomes a desert far more bereft of life than the Sahara. This is because at temperature above about 15°C the nutrients in the ocean surface are rapidly eaten and the dead bodies and detritus sink to the regions below. There is plenty of food in the lower waters, but it cannot rise to the surface because the cooler lower ocean water is denser than water at the surface ... This is important because ... Earth is a water planet with nearly three-quarters of its surface covered by oceans. Life on land depends on the supply of certain essential elements such as sulfur, selenium, iodine and others. Just now these are supplied by ocean surface life as gases like dimethyl sulfide and methyl iodide. The loss of this surface life due to the heating of these waters would be catastrophic [28].*

Rising ocean surface temperature will lead to catastrophic decline of both ocean surface life and land life.

How has Gaia, the Earth system, maintained its temperature within a homeostatic range: Lovelock suggests the following mechanism as a working hypothesis:

> *"In modern times, carbon dioxide is a mere trace gas in the atmosphere compared with its dominance on the other terrestrial planets or with the abundant gases of Earth, oxygen and nitrogen. Carbon dioxide is at a bare 340 parts per million by volume now. The early Earth when life began is likely to have 1000 times as much carbon dioxide... As the Sun warmed, two processes took place. The first was an increase in the rate of evaporation of water from the sea and, hence, rainfall; the second, an increase in the rate of the reaction of carbon dioxide with the rocks. Together, these processes would increase the rate of weathering of the rocks and so decrease the carbon dioxide. The net effect would be a negative feedback on the temperature rise as the solar output increased... [Lovelock then added a third process involving living organisms]... living organisms act like a giant pump. They continuously remove carbon dioxide from the air and conduct it deep into the soil where it can react with the rock particles and be removed [29].*
> *"... If confirmed, it suggests that cloud cover and low carbon dioxide operated in synchrony as part of a geo-physiological process to keep the Earth cool... [30].*
> *"From the very beginning of life on Earth, carbon dioxide has had a contradictory role. It is the food of photo-synthesizers and therefore of all life; the medium through which the energy of sunlight is transformed into living matter. At the same time, it has served as the blanket that kept the Earth warm when the Sun was cool. A blanket that, now that the Sun is hot, is becoming thin; yet one that must be worn, for it is also our sustenance as food. We have seen earlier how the biota everywhere on the land and sea are acting to pump carbon dioxide from the air so that the carbon dioxide which leaks into the atmosphere from volcanoes does not smother us. Without this never-ceasing pumping, the gas would rise in concentration within a million years to levels that would make the Earth a torrid place and unfit for almost all life here now. Carbon dioxide is like salt. We cannot live without it, but too much is a poison" [31].*

The details of the working hypothesis may yet to be worked out. But two takeaways are sufficiently clear and they are: (Surmise-1) the necessity to keep the Earth system cool in order to keep it within the temperature *homeostatic ranges*—while keeping in mind of other important *homeostatic ranges* of the *metric-set*; (Surmise-2) carbon dioxide is the critical element involved in the mechanisms of achieving the goal.

This brings us to the two metrics of ***homeostatic state***, i.e., underlying all the homeostatic ranges of Surmise-1 is the idea of keeping the Earth system safely from thermodynamic equilibrium—corresponding to Surmise-2, in which carbon dioxide is the proxy of *entropy difference of the system from system-at-reference-equilibrium-state*. This is why it is necessary to abandon the conventional thermodynamics, in which the idea would be a nonstarter, to embracing, instead, a new engineering-thermodynamics. Only with the second law as both a principle of inevitable entropy growth and a principle of entropy growth potential, it is possible to keep the Earth system safely from thermodynamic equilibrium.

One example of solutions for the goal is the electrification of space heating. The purpose of the essay is not to outline such kind of specific solutions but to use such a solution-example to advance the argument that such opportunities exist only if we frame the crisis and problem in systems-framework in terms of EGP management.

In this systems-thinking framework, we do have an existential threat. The threat is, however, not the threat of running out of fossil fuel or fossil energy. The standard narrative of such kind of thinking is that we have abundant solar energy and the solution to our problem is to find ways of converting a small part of solar energy (including wind energy) into useful energy. This is clearly the wrong way to look at the problem. If sunlight is our savior (which is) in this sense, a warming Sun should have been a welcoming development, in opposite to the idea of *heat threat* from a hot Sun [32].

Transition from fossil energy to renewables *is* a good idea, not because we welcome a warming Sun as a source of heat energy. But because the solar output received by Earth is a "form of entropy flow of very low value." Assuming the Earth system is in a state of energy balance, the Earth infrared radiative heat outflow equals the solar radiative heat inflow received by the Earth. The corresponding values of entropy flow received by the Earth from the Sun and of entropy out-flow from the Earth to out-space will be significantly very low and very high, respectively. That means that very large entropy growth potential exists in the difference of the two flows.

That means that large opportunities exist in the management of entropy growth potential. Some of those opportunities, such as electrification of heating, can be related to the control of carbon dioxide. That also means that while a warming Sun poses heat threat to the Earth, it also presents greater opportunities for EGP management.

6. Conclusion

Humans has experienced energy transitions throughout its history and the current transition from fossil energy to renewable energy is the latest example. But this latest example is different: this energy transition results from the threat of global warming—which is generally attributed to the short-term increasing of carbon dioxide in the atmosphere but also to the long-term heat threat posed by a warming Sun, according to the Gaia theory. Appreciation of the nature of this combination of proximate cause and ultimate cause necessitates for us to take a systems-thinking about the Earth system as a whole. Energy transition to renewable energy is certainly correct, especially reassuring since solar energy received by the Earth is 6,900 times of the energy needs of humans. The solution would be then how to convert a small part of which into useful forms for human consumption.

But justification of such a step in the narrow terms of energy is wrong. Humans face existential threat of global warming as *heat threat* from the Sun, not as energy threat of running out of fossil fuels. Solving Earth's heat threat necessitates us to take consideration of its proximate and ultimate causes with systems-thinking framework in terms of the management of EGP. Only by taking this perspective, we can address the root-issue of the heat threat—as well as seeing a warming Sun as both threat and opportunity. One of the opportunities is electrification of space heating, a paradigmatic example of systems solutions. Other possible solutions may be formulated by taking systems-thinking in terms of the management of EGP that may address some of humans' *Existential Risk and the Future of Humanity* [2].

Author details

Lin-Shu Wang* and Peng Shi
Stony Brook University, Stony Brook, NY, USA

*Address all correspondence to: lin-shu.wang@stonybrook.edu

References

[1] Smil, Vaclav. 2010. Energy Transitions. History, Requirements, Prospects. (Praeger)

[2] Toby Ord, *The Precipice: Existential Risk and the Future of Humanity*, Hachette, 2020

[3] L-S. Wang (2021). "Science in postmodernity: Saving science from scientism," *Science and Education* (submitted Apr. 1, 2021_SCED-S-21-00130)

[4] Ted Poston. "Foundationalism," Internet Encyclopedia of Philosophy.

[5] Tzuchien Tho (2020). "Mechanical Philosophy: Reductionism and Foundationalism," *Encyclopedia of Early Modern Philosophy and the Sciences* (edited by D. Jalobeanu, C. T. Wolfe). Sect. M, pp.1-7

[6] DC Phillips and NC Burbules (2000). *Postpositivism and Educational Research* (Rowman & Littlefield), pp. 5-26.

[7] Laplace's Demon, Retrieved April 3, 2021, from Information Philosopher–Web site https://informationphilosopher.com/freedom/laplaces_demon.html

[8] Papineau, David. "Naturalism", *The Stanford Encyclopedia of Philosophy* (Summer 2020 Edition), Edward N. Zalta (ed.), URL = https://plato.stanford.edu/archives/sum2020/entries/naturalism/

[9] "Efficiency movement," *Wikipedia. Efficiency movement - Wikipedia.*

[10] Erik Brynjolfsson and Andrew McAfee (2014). *The Second Machine Age: Work, Progress, and Prosperity in a Time of Brilliant Technologies.* (Norton)

[11] Emden, *Nature 141*: 908–909 (1938)

[12] Peng Shi (2021) Analyses on Promising End-user Side HVAC Technologies for Natural Gas, Electricity and Control, *doctoral dissertation, Stony Brook University* (June, 2021)

[13] New York Independent System Operator, 2019, "Real-Time Fuel Mix, https://www.nyiso.com/real-time-dashboard"

[14] W. Thomson, *Mathematical and Physical Papers of William Thomson*, **1** (Cambridge University Press, 1911): pp. 1-571. (the paper page numbers are 511-514).

[15] M. Planck (1969). *Treatise on Thermodynamics,* 3rd edition (Dover, New York)

[16] L-S. Wang (2020). *A Treatise of Heat and Energy* (Springer, Cham, Switzerland)

[17] L-S. Wang and P. Shi (2020). "Against the energy-conversion-doctrine: Why energy conservation manifests universal connection rather than universal interconversion?" *Renewable Energy—EnerarXiv*

[18] H. Poincaré, *Science and Hypothesis*, (The Science Press, Lancaster, PA, 1913), pp. 122-123.

[19] L-S. Wang (2021). "Progress in entropy principle," to appear in *International Journal of Design & Nature and Ecodynamics (***IJDNE***)*

[20] Lovelock JE (1972) Gaia as seen through the atmosphere. Atmospheric Environment 6: 579–580.

[21] Lovelock JE and Margulis LM (1974a) Atmospheric homeostasis by and for the biosphere: The Gaia hypothesis. Tellus 26: 2–10.

[22] Lovelock JE and Margulis LM (1974b) Homeostatic tendencies of the Earth's atmosphere. Origins of Life 5: 93–103.

[23] Margulis L and Lovelock JE (1974) Biological modulation of the Earth's atmosphere. Icarus 21: 471–489.

[24] Dawkins R (1982) The Extended Phenotype. Oxford: Oxford University Press.

[25] Doolittle WF (1981) Is nature really motherly? The Coevolution Quarterly 29: 58–63.

[26] Timothy M Lenton et al. (2020) "Life on Earth is hard to spot," The Anthropocene Review 2020, Vol. 7(3) 248–272

[27] Doolittle WF (2019) Making evolutionary sense of Gaia. Trends in Ecology & Evolution 34: 889–894.

[28] Lovelock JE (2019). *Novacene* (Allen Lane), p.60

[29] Lovelock JE (1988). *The Age of Gaia* (Norton & Company). pp. 126-127

[30] *The Age of Gaia*, p.142

[31] *The Age of Gaia*, p.148

[32] *Novacene*, pp. 57-66

Chapter 2

A Comprehensive Review on Available/Existing Renewable Energy Systems in Malaysia and Comparison of Their Capability of Electricity Generation in Malaysia

Hadi Nabipour Afrouzi, Yuhani Pamodha Wimalaratna, Jubaer Ahmed, Kamyar Mehranzamir, San Chuin Liew, Chin-Leong Wooi and Bazlul Mobin Siddiquea

Abstract

Malaysia is one of the fastest emerging and developing countries in the world. To drive the economical workhorse, large amounts of power is required. The power demand has risen to 156,003 GWh per year in the year 2016, almost 30,000 GWh more than 5 years prior. Fossil fuels such as natural gas, coal, oil, and diesel have been the driving force powering Malaysia's grids. However, these resources will not last forever, and they do harm to our environment. To counter this, renewable energy (RE) projects have been constructed all around Malaysia. This paper discusses on available and existing renewable energy systems (single/hybrid) in Malaysia and provides a comparison of their electricity generation capabilities. The renewable energy sources that are covered in this paper include Solar, Hydropower, Biomass, Tidal and Geothermal. At the moment, hydropower is the largest renewable energy producer, contributing to almost 15% of the country's total energy generation. A lot of resources have been channeled towards the initiative of hydropower and it has definitely borne much fruit. This is followed by Solar Energy. Even though it is not as successful as hydropower, there is still a lot of avenues for it to grow in a tropical country like this. Malaysia is still relatively new in terms of power generation using biomass sources. There has been a gradual increase in the power generation using biofuels through the years and its future does look bright. Energy generation from wind, tidal, and geothermal sources has been rather challenging. Because of Malaysia's geographical location, it experiences slow winds on average throughout the year. This has led to insufficient output for its financial input. Besides that, Malaysia also has relatively low tide, if compared to other Asian countries such as Indonesia and the Philippines. This contributed to the failure of tidal energy in Malaysia, but there have been signs of locations that can be suitable for this energy generation. Besides that, the country's first geothermal power plant project failed due to a lack of preparation and discipline during the project's execution. There is a high initial cost for geothermal projects, and the chances of failure are high if the necessary precautions are not followed. This could

IntechOpen

be one of the reasons why this branch of renewable energy has not been explored deeply.

Keywords: Malaysia, renewable energy, solar energy, hydropower, biomass, geothermal

1. Introduction to renewable energies

The demand for electricity increases new ways of electrical generation are required that is both cleaner and safer. In Malaysia, research has shown that about 3.8% of the population reside below the poverty line. Most of these people are located in rural areas in Malaysia. The electricity coverage in Peninsular Malaysia is at 99.62%, while Sabah and Sarawak's electrical coverage is around 79%. The challenge is to build a grid system through jungles and mountain. To add to that, building a grid system through these types of the area will also not be economic. A way to solve the problem would be through the implementation of Renewable Energy (RE) in these villages. The main source of Malaysia's energy supplies is from Natural Gas, Hydro, Oil, RE and Coal. Among these five Energy Sources (ES), coal supplies the most energy in terms of electricity production at 26,177 GWh. The types of RE which were researched are Solar, Wind and Hydropower. Not all of these energies are widely used in Malaysia. Some Renewable Energy Sources (RES) are ideal because of the terrain or weather in Malaysia, while others are under the research phase to determine the possibility of implementation in Malaysia. Malaysia has been amply endowed RESs such as Solar and Biomass Energy. However, these ESs have been greatly underutilized. A comparison of each REs was done to understanding the applicability of each of these resources in the Malaysian context. Being able to implement these ESs especially in areas not connected to the national grid would be beneficial to Malaysia to improve the living quality of Malaysians in rural areas.

2. Malaysia solar energy information

2.1 Introduction to solar energy

Sun is the ultimate resource on earth as it is responsible for all the weather conditions and ESs on earth. Sun emits Solar Energy due to the nuclear fusion reactions in the sun's core and subsequently produces a tremendous amount of energy. However, small portion of it is directed towards earth in the form of light and heat. Solar energy, which correlates to the sunlight's photons has an abundant potential that can fill our global needs if it is harnessed in the right way.

Generally, there are two main ways to harness Solar Energy, which is using either photovoltaics or solar thermal collectors. Photovoltaic (PV) or commonly known as solar cells, comes in various shapes and are made from electricity producing materials such monocrystalline silicon, polycrystalline silicon and thin film solar cells. When sunlight gets in contact with the solar cells' semiconductor material, they get absorbed and consequently, generate electricity [1]. This conversion is mainly due to the photovoltaic effect. When this effect occurs, the photons from the sun's radiation knocks electrons loose, causing them to flow and thus generate electricity. The initial generated current is the direct current (DC). In order for, this can be stored in the battery and used for DC appliances. To make it useable for regular households, it is first converted to alternating current (AC) using an inverter. If the system is connected to the grid, then additional electricity is fed to the main supply. The other way of tapping the sun's energy is by capturing the heat produced by the solar radiation. This form of harnessing is usually done in a large scale in such a fashion that power stations are built. These power stations are called Concentrated Solar Power (CSP) plants. The term concentrated comes from a large number of mirrors in the plant which are used to focus the sun's rays on tubes containing molten fluid that can store heat well. The molten fluid then is used to convert water into steam. Subsequently, the steam produced rotates a turbine and thus, generate electricity [2].

In Malaysia, solar cells are commonly used to generate electricity. In 2018 alone, 467344.2 MWh of power was generated based on Malaysia's Feed-in Tariff (FiT) system. Comparing this figure to the other RE, harnessing solar energy comes up on top. Although the nation is exposed to long hours of sunlight daily, the average maximum amount of energy produced per solar cell has an efficiency of 15–20 percent. This efficiency poses an issue whereby not many investors would invest in the technology. A way to overcome this situation in Malaysia is by constructing Large Scale Solar (LSS) power plants. This way, the amount of electricity generated can be maximized. The LSS power plants are not to be mistaken with CSP plants. The main difference between the two is that LSS captures light via solar cells and converting them into electricity whereas CSP captures heat which is transformed into mechanical energy that rotates a turbine and subsequently produces electricity. In Malaysia, CSPs are not developed yet. Having the minimal Direct Normal Irradiance (DNI) within the range of 1900 to 2000 kWh/m2/year, is the main requirement to start a CSP project. However, the DNI for Malaysia is below this threshold and it is due to the geographical position of the country which is not situated in high solar insolation zones [2].

2.2 Solar energy in Malaysia

2.2.1 Pajam, Negeri Sembilan

On 20th March 2012, an 8 MW large scale solar PV plant, developed by Cypark Resources Berhad, was officially launched and operational. **Figure 1** shows the aerial view of the solar power plant. This project is the first-ever completed LSS above 1 MW and operational under the Sustainable Energy Development Authority's (SEDA) Feed-In-Tariff Mechanism (FiT) in Malaysia. The LSS has received two accolades by the Malaysia Book of Records as it is recognized as one of the largest grids connected solar parks in the nation. The land coverage by the LSS is approximately 41.73 acres and it is equipped with 31, 824 solar panels [3]. Being the first of its kind in the country, the RM150 million project has the ability to power

(a) (b) (c)

Figure 1.
Aerial view of three major Solar Energy Projects in Malaysia. (a) An aerial view of the 8 MW large scale solar photovoltaic plant [8], (b) An aerial view of Mukim Tanjung 12 Solar Photovoltaic Plant [9] and (c) An aerial view of Sungai Siput solar photovoltaic power plant [13].

over 17,000 households annually. In 21 years from its initiation, it is expected to generate up to RM500 million worth of electricity. This is equivalent to the power generated by 9, 300 tons of coal each year. For the environment, it is capable of reducing 14, 335 tons of carbon emissions and 664 tons of methane gas annually [3, 4].

2.2.2 Mukim Tanjung 12, Kuala Selangat, Selangor

As of November 2018, the nation's largest LSS has started its operation. The project won in competitive bidding by Tenaga Nasional Berhard (TNB) and subsequently, the project started its development in July 2017. The 10 km of 132 kV power and fiber optic underground cables were connected to 230, 000 solar panels in this plant. This LSS is capable of producing 50 MW of electricity to the national grid. The total cost of this project is approximately RM339 million. The total land size used is up to 242.16 acres. Due to the success of TNB, this project serves as a booster and aspiration in further developing more RE projects in Malaysia. Consequently, by 2030, Energy, Science, Technology, Environment and Climate Change Ministry has set a goal to increase the country's electricity usage powered by 20% based on Res [5] Also, TNB's success in this project has led the company to secure RM144 million in developing a second large scale solar project for the country [5, 6] **Figure 1** shows the aerial view of the LSS.

2.2.3 Sungai Siput, Perak

On 27 November 2018, one of Malaysia's advance solar PV power plant has started its operation. This project began on 16 March 2017 when Sinar Kamiri Sdn Bhd signed a power purchase agreement with Tenaga Nasional Berhad to develop and operate a 49 MW large scale solar photovoltaic power plant which cost around RM270 million [6]. This LSS is situated in Sungai Siput Perak over a land size of 150 acres and is equipped with 170, 961 panels. In this land, the complex mountain topography has posed many challenges for the developers as this would often cause shadows, string mismatches as well as high temperature and humidity. However, the land offers a long duration of sunshine and high solar irradiance throughout the year. To overcome this topographical situation, the developers have integrated Huawei Fusion Solar Smart PV Solution into the grid [7]. This includes the smart PV string inverter SUN2000-42KTL to troubleshoot the string mismatch issues faced in Sungai Siput as well as a PLC technology which helps to deliver a simpler system with safer and more reliable data transmission. As a consequence of using these PV systems, this LSS has obtained 2% higher energy yields and a 50% increase in efficiency compared to other LSS of the same scale. **Figure 1** illustrates the aerial view of the LSS.

2.2.4 Kudat, Sabah

In July 2017, RM250 million green socially responsible investment (SRI) sukuk has been issued to Tadau Energy Berhad to further develop the current state of RE usage in Malaysia. Also, this green SRI sukuk receives funds for the projects due to its international endorsement and potential tax benefits via deduction of issuing expenses against the taxable income of the issuer [8]. In other words, it helps companies to achieve their corporate social responsibilities. With this cash in hand, the company started to develop a 50 MW solar power plant in Kudat, Sabah which covers up to 189 acres. This LSS is equipped with 188, 512 solar panels. 2 MW out of 50 MW of the available power channels the local Kudat electricity grid while the remaining 48 MW is channeled into 132 kV transmission line which is distributed through Sabah [9].

2.3 Comparing the projects

Based on the information in **Table 1**, the selected 4 LSS projects can be arranged and compared between one another in terms of generation capacity, number of solar panels, generation capacity per solar panel, land area and project cost.

First, among the projects, the LSS which has the highest generation capacities are Mukim Tanjung 12, Sungai Siput and Kudat. Each of these LSS has a generation capacity of around 50 MW. Subsequently, this is followed by Pajam at 8 MW. The difference in the capacities is based on the purpose of the project. For example, the reason why Kudat's generation capacity is high is that the power generated is used to supply the local villages as the remaining is supplied to the power grid, which is then distributed throughout Sabah [9]. A smaller LSS may not have the same purpose and the demand for electricity in the area may not be as high as areas with more population or activities. Alongside this reasoning, it corresponds as well with the total number of panels. Although Mukim Tanjung 12, Sungai Siput and Kudat have similar generation capacity and project motives, the number of panels used at each plant is different. As seen in the table, Mukim Tanjung 12 uses 230, 000 panels, Sungai Siput uses 170, 961 panels and Kudat uses 188, 512 panels. Coinciding with this information, it can be inferred that the panels which are used in each power plant have different efficiencies. For instance, although Sungai Siput uses fewer solar panels compared to Mukim Tanjung 12 nad Kudat, it is still having a similar power output as the other two. This is because the panels which the LSS has, uses Huawei's Fusion Solar Smart PV Solution panels. These panels have the potential to increase energy yields, maximize the return of investments (ROI) and helps customers optimize initial investments. Also, DC combiners are not needed in these plants [7, 8]. The reasoning in the paragraph is also backed up by the amount of each that each panel can generate. From the table based on the third row, the highest yielding panel to lowest is Sungai Siput at 286.61 W/panel, Kudat at 265.23 W/panel, Pajam at 251.31 W/panel then Mukim Tanjung 12 at 217.39 W/panel. From here, the quality of the panels used in both Sungai Siput and Kudat are of higher efficiency. Next, the amount of land size used from highest to lowest is in an order of Mukim Tanjung 12 at242.16 acres, Kudat at189 acres, Sungai Siput at 150 acres and lastly Pajam at 41.73 acres. The land coverage is closely dependent on the required generation capacity as well as the yield per panel. If the required generation capacity is low, the land size will not cover over a large area as seen in Pajam.

Location	Pajam, Negeri Sembilan	Mukim Tanjung 12, Selangor	Sungai Siput, Perak	Kudat, Sabah
Generation Capacity (MW)	8	50	49	50
Number of Solar Panels	31, 824	230, 000	170, 961	188, 512
Yield per Solar Panel (W/panel)	251.38	217.39	286.61	265.23
Land Area (acres)	41.73	242.16	150	189
Project Cost (RM in millions)	150	339	270	250

Table 1.
Summary of information of the four selected LSS.

Also, if the yield per panel is high, the land size needed is small. The last aspect that can be compared is the project cost. This correlates with the land size, efficiency of the panels and ease of installation. Generally, it would cost more for a land of a bigger size. This goes the same for a higher quality panel. In terms of ease of installation, it depends on the safety factors that are given to each component in the power plant based on the land's topography and weather. Some areas could be flat land while some are covered by hills. In Mukim Tanjung 12, since its land size is large, it accounts for the high cost of the project. Subsequently, Sungai Siput's project cost is relatively high as well and this is due to the hilly area which the LSS is built on as well as the quality of the solar panels.

2.4 Comparing solar energy in Malaysia to the World

In Malaysia, the country's first LSS was developed in 2012. However, it does a gradual impact on the awareness of people on using RESs. Ever since then, more and more LSS projects have been developed and the country has started to see this RE's advantages. Consequently, in the current years, the country has new policies such as the Renewable Energy Transition Roadmap (RETR) 2035 which aims to further explore the possible strategies and action plans to reach the country's renewable target of 20% in the national power mic by 2025. In **Table 2**, the new addition in the panels is around the same value for each year. This is due to the country's reliance on coal and fuel which has also been one of the main sources of the country's economy. As a result, transitioning to another form of ES requires confidence built up by the country. Nonetheless, this issue is slowly alleviated as the awareness of using more RESs has increased every year.

Among these four countries, in terms of annual production in 2018 (**Figure 2**), China has produced the most energy of a figure close to 80 GW. This is followed by Malaysia with an annual production of approximately 15 GW. Subsequently, Japan produced around 5 GW in that year and lastly, the USA has produced half of Japan's [12].

2.5 Section conclusion

Based on the trends, the number of solar PV additions by each of the countries has plateaued in recent years due to hurdles faced within the country. Considering this, some other countries have been growing in this field including Malaysia. Ever since the first LSS was developed, the country has been developing more projects that is able to harness the sun's energy. Due to the country's accumulative efforts, Malaysia has the potential to become one of the leading countries in solar PV generation given that further research and development is given into this field.

3. Malaysia hydropower information

3.1 Introduction to hydropower

Hydropower is the conversion of Kinetic Energy (KE) of water into electricity and is considered a RES due to the water cycle being constantly renewed by the sun. According to an article on the US Geological Survey website, a hydropower dam works by having the water in the reservoir flow into a pathway called the penstock when the sluice gate is opened. This penstock is directly connected to the turbine, which is spun by the force of the water moving from a location of higher pressure to one of lower pressure. The water itself then flows out to a river below, whereas the turning of the turbine causes the conversion of KE from the water force into

Country	2015		2016		2017		2018	
	Total solar PV Addition (MW)	Net Solar PV Capacity (MW)	Total solar PV Addition (MW)	Net Solar PV Capacity (MW)	Total solar PV Addition (MW)	Net Solar PV Capacity (MW)	Total solar PV Addition (MW)	Net Solar PV Capacity (MW)
China	15,150	43,530	34,540	78,070	53,000	131,000	44,018	175,018
US	7,300	25,620	14,730	33,100	8,173	41,273	8,419	49,692
Japan	11,000	34,410	8,600	42,750	7,000	49,000	6,500	55,500
Malaysia	63	231	54	286	50	386	52	438

Table 2.
Net solar PV capacity and total addition by country from 2015 to 2018 [10, 11].

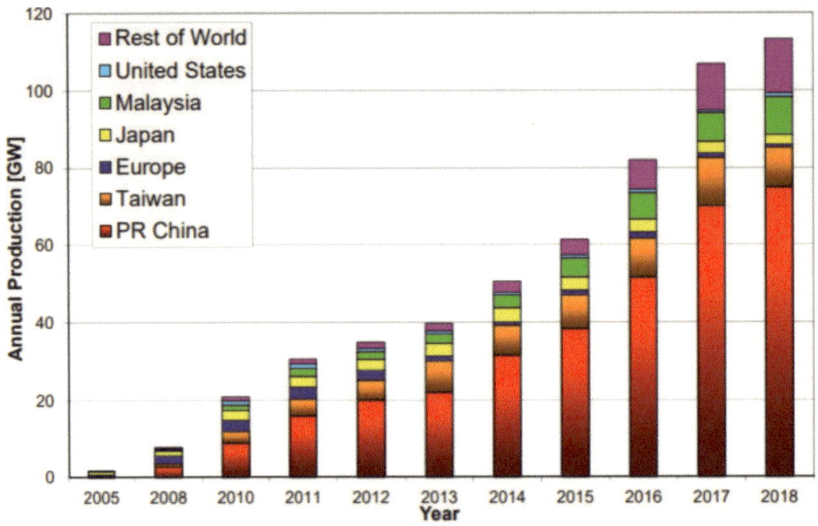

Figure 2.
Global PV cell production from 2015 to 2018 [12].

mechanical energy for use by the generator, which is connected to it by way of shafts or gears. This turning of the turbine also causes the rotor within the generator to turn and consequently causes the electromagnets on its edge to move past the stators placed in a static position outside the rotor, allowing for the conversion of the mechanical energy from the turbine into Electrical Energy (EE). The electricity produced from this conversion process is then carried out to other locations and facilities by way of power to transmission lines connected directly to the generator.

3.2 Hydropower in Malaysia

International Hydropower Association states that the installed hydropower capacity is 6094 MW in 2016, with hydropower generating roughly 11% of the country's electricity and less than 20% of the technically feasible generation potential utilized to date in their article from May 2017 [14]. The following comparison has been done by choosing the most five powerful Hydropower plants in Malaysia.

3.2.1 Bakun Hydroelectric Plant, Sarawak

The Bakun Hydroelectric Plant is located on Batang Lui in the upper parts of the Rajang River, roughly 37 km upstream of the town of Belaga in Sarawak. The plant is powered by eight 300 MW turbines, allowing for an installed generation capacity of 2400 MW and has a power transmission system that directly connects to the existing power transmission network in Sarawak. The plant has been operational since 2011 and produces an average electricity generation of 1700 MW to 2110 MW depending on demands. The dam is considered to be the largest and tallest Concrete-Faced Rock-fill dam in South East Asia with a 205 m height and 750 m length, with the capability to contain 16.93 million m^3 of water, allowing the reservoir a surface area of around 695km^2 with a catchment area of 14750 km^2 [13].

3.2.2 Murum Hydroelectric Plant, Sarawak

The Murum Hydroelectric Plant was completed back in 2016 and is located on the Murum River in the upper region of the Rajang River Basin, roughly 200 km from Bintulu. The plant is powered by four 236 MW turbines, which totals to an installed generation capacity of 944 MW [13, 14], with its average production being around 635 MW and would be delivered through the state power grid. The dam is 141 m high and 473 m long, with a reservoir area of $270km^2$ and a catchment area of $2750km^2$. The cost of the project totaled about RM 4.8 billion [15]. Besides that, Murum also has the world's tallest stepped chute spillway that helps to reduce KE by aerating the water overflow, which also helps to preserve the riverine ecosystem and the Batu Tungun rock formation, which is considered sacred to the local Penan community [14].

3.2.3 Pergau Hydroelectric Station, Kelantan

The Pergau Hydroelectric Station is located on the Pergau Lake, around 100 km away from Kota Bharu, Kelantan. The plant is powered by four 150 MW turbines totaling 600 MW of installed generation capacity and was designed to operate at a daily load factor of 25%. The Kuala Yong dam, which the power station receives its water from, is 75 m high [15, 16], with a $54km^2$ upper catchment and lower plain area. Besides that, the station also has a $1.5 \times 10^6 m^3$ reregulating pond designed to accept peak generation flows and to release them into the river in a controlled manner, which also had a generating cycle of 5.5 hours when at full station output. The cost of the project is totaling to RM 2.23 billion [17].

3.2.4 Sultan Mahmud Power Station, Terengganu

The Sultan Mahmud Power Station was completed in 1985 and located 55 km southwest of Kuala Terengganu on the Kenyir Lake. The plant is powered by four 100 MW turbines, totaling 400 MW of installed generation capacity, with continuous generation being 165 MW. The dam is 155 m high and 800 m long in crest, with a reservoir area of $369km^2$ and a catchment area of $1260km^2$. The water height is around 120 m at minimum capacity and can go up to 153 m when it's at maximum, with a full supply level of 145 m [16, 17]. The lake itself can store 13.6 billion m^2 of water, with its deepest point being 145 m. Besides that, it can also release any excess water flow in the reservoir directly downstream into the Terengganu River.

3.2.5 Ulu Jelai Hydroelectric Power Plant, Pahang

The Ulu Jelai Hydroelectric Power Plant was completed in 2016 and is located in the Cameron Highlands, Pahang on the Bertam River. The plant is powered by two 186 MW turbines which are placed in an underground plant [18], totaling to 372 MW of installed generation capacity for electricity. The Susu Dam, which is the dam that forms the Susu Reservoir of this hydropower plant, was built using almost $750{,}000m^2$ of concrete through the Roller-Compacted Concrete (RCC) method, a very modern way to build such a dam. Said dam is measured to be 88 m high and 460 m long in the crest, with a $0.1km^2$ catchment area. The total cost of the project was RM 4.2 billion and is expected to reduce 250,000 tons of carbon dioxide equivalent per year by substituting coal or fossil fuel-based generator stations during peak hours, according to a United Nations report [17, 18].

3.3 Comparing the projects

Based on the information gathered here, the five selected hydropower projects can be arranged and compared between each other within the categories of installed generation capacity, dam size, catchment area, reservoir area, and project cost (**Table 3**) [15, 18].

When it comes to total installed generation capacity, the Bakun has the highest of the five at 2400 MW, followed by Murum at 944 MW, Pergau/Sultan Ismail Petra at 600 MW, Sultan Mahmud at 400 MW and lastly Ulu Jelai at 372 MW. The reason behind the high output behind Bakun is not only it's high number of turbines, but the capacity of 300 MW that each turbine is capable of, which in and off itself is close to rivaling the entire output of the Ulu Jelai station at 372 MW. This makes it the most powerful hydroelectric power plant in Malaysia and the largest power generation facility in Sarawak, as it also supports the Sarawak Corridor of Renewable Energy (SCORE) initiative required for the energy-intensive heavy industries such as the Samajaya Industry Park. In terms of the dam sizes, Bakun has the largest at 205 m high and 800 m long, followed by Sultan Mahmud at 155 m high and 800 m long, Murum at 141 m high and 473 m long, and lastly Ulu Jerai at 88 m high and 460 m long. The length of the Pergau/Sultan Ismail Petra was not given but can be assumed to be the smallest of the five as the height is only 75 m. As for catchment area, which is a land area where water can flow into the plant reservoir [15, 18], the largest is Bakun at 14750km^2, followed by Murum at 2750km^2, Sultan Mahmud at 1260km^2, Pergau at 54km^2 and lastly Ulu Jerai at 0.1km^2. Ulu Jerai is the smallest of the bunch as it uses the Bertam River, whereas the rest have a larger area to work with as they are built on lakes and other large bodies of water. Out of the five stations, only Bakun, Murum and Sultan Mahmud have a listed reservoir size at 695km^2, 270km^2 and 369km^2 respectively. Having a reservoir allows for the storage of water as conversion fuel for a later date [17], meaning that Bakun has the largest water reserve of them all and thus can use more water to generate more electricity in comparison. When it comes to cost, the most expensive project was Bakun at RM 7.3 billion, Murum at RM 4.8 billion, Ulu Jelai at RM 4.2 billion, and lastly Pergau at RM 2.23 billion, whereas the cost for production of the Sultan Mahmud plant was nowhere to be found but could be assumed to be between Pergau and Murum due to the size being between those two and that it was completed in 1985. The cost of Bakun being the highest is because it uses more turbines that are very powerful in order to produce more power than the rest of the ones on the list combined at peak usage, not to mention the size of the construction project itself.

To wrap this part up, the Bakun Hydroelectric Plant is the best hydroelectric plant available in Malaysia due to the amount of installed generation capacity for electricity that it can provide due to the massive size of the project itself, but such power comes at a great price tag. In comparison, a project such as the Sultan Ismail Power Station or Pergau Hydroelectric Station would be a more feasible one to create in a higher quantity for a developing country such as this due to the lower

Hydropower Station	Max Power (MW)	Turbine Amount	Dam Dimensions (h x l)	Surface Area (km^2)	Catchment Area (km^2)	Project Cost ($ Billion)
Bakun	2400	8 x 300	205 x 750	695	14750	7.3
Murum	944	4 x 236	141 x 473	270	2750	4.8
Pergau	600	4 x 150	75 x —	54	—	2.23
Sultan Mahmud	400	4 x 100	155 x 800	369	1260	—
Ulu Jelai	372	2 x 186	88 x 460	—	0.1	4.2

Table 3.
The data for each of the five selected hydroelectric power plants.

costs and less space requirement, while still capable of pumping out a respectable amount of electricity for the towns, villages and cities found in this country.

3.4 Comparing hydropower in Malaysia to the World

Malaysia's annual hydropower energy production is rated at 4.5 Mtoe/year (Megaton of energy per year) [16] with an installed hydropower capacity of 6094 MW and a hydropower usage percentage of 11%. The largest dam or hydropower facility in Malaysia is the Bakun Dam at 2400 MW of installed generating capacity.

From the information gathered here, the leading hydropower nations were China, Brazil, and Canada, with 96.9 Mtoe/year, 32.9 Mtoe/year and 32.3 Mtoe/year (megaton of energy per year) respectively, whereas overall energy input from hydropower in Malaysia is totalled to be 4.5 Mtoe/year (kiloton of energy per year). In comparison to the 3 countries stated previously, Malaysia's hydropower energy input is extremely tiny in comparison (**Table 4**). When it comes to the respective strongest dams in terms of output, China's Three Gorges Dam is the highest at 22.5GW of installed generating capacity, followed by the joint Brazil/Paraguay Itaipu Dam at 14GW and lastly Canada's Robert-Bourassa Dam at 5616 MW. Malaysia's largest one, Bakun at 2400 MW, is respectable and considered the largest in South-East Asia [13, 17], but the output is nothing compared to these giants. It is generally more than enough for providing electricity to Sarawak itself and can support the local heavy industries found in the state. When it comes to percentages of hydropower usage for electricity in a country, Brazil has the largest at 64%, followed by Canada at 62%, China at 20%, and lastly Malaysia at 11%. Malaysia's percentage of hydropower usage is lower compared to these other countries as Malaysia still relies heavily on coal for most of their power stations [15], which should be changed as soon as possible as it's a non-RES and could dry up in the future. In terms of installed hydropower capacity, China is the largest at 341,190 MW, followed by Brazil at 100,273 MW, Canada at 79,323 MW and lastly Malaysia at 6094 MW. Considering how small Malaysia is compared to all the three countries above, it's understandable that the installed hydropower capacity is far lower than them as Malaysia has much less space and water bodies to work with in comparison, while not to mention Malaysia's economy not being as strong as them (**Figure 3**).

3.5 Section conclusion

Taking an example out of Canada or Brazil here would be a good idea as there are many rivers and water bodies that could be exploited for hydroelectricity generation, however building more mega-dams like Bakun could harm the rainforests

Country	Annual Energy (Mtoe/year)	Strongest dam & output (MW)	Installed hydropower capacity (MW)	Usage Percentage (%)
China	96.9	Three Gorges' Dam (22,500)	341,190	20
Brazil	32.9	Itaipu Dam (14,000)	100,273	64
Canada	32.3	Robert-Bourassa Dam (5,616)	79,323	62
Malaysia	4.5	Bakun Dam (2,400)	6,094	11

Table 4.
Data for three of the world's leading hydropower countries in comparison to Malaysia's.

Figure 3.
Hydropower Major projects in Malaysia. Aerial view of the (a) Bakun Hydropower, (b) Murum Hydropower, (c) Yong Hydropower and (d) Kenyir Hydropower.

and animal species that makeup country's ecosystem, not to mention potentially displacing the natives and eating large sums of money that could be used for other equally beneficial projects. Therefore, a balanced method of implementing hydropower while maintaining the ecosystem should be explored, so that Malaysia could progress to the future with hydropower while still maintaining the well-being of Malaysia's unique ecosystem.

4. Malaysia biomass energy information

4.1 Introduction to biomass energy

Biomass is a type of fuel developed from organic materials. It is both sustainable and renewable in terms of generating EE. The organic materials are obtained from living and recently living things. These materials can include scraps of lumber, manure and forest debris. Biomass power is able to generate electricity which is carbon neutral through renewable organic waste [19]. This energy releases heat when burnt. These energies are utilized through burning them to produce steam to run turbines which in turn generates electricity.

4.2 Biomass in Malaysia

4.2.1 Biomass resources in Malaysia

Malaysia is a country filled with many conventional energy resources. These energy sources include oil and gas which are non-renewable and RESs like solar, hydro and biomass energy. For biomass, Malaysia has plenty of opportunities as far as exploiting biomass energy in Malaysia. Malaysia is filled with agricultural biomass and wood waste which can be exploited and used to replace non –RESs in use [19].

4.2.2 Palm oil biomass

Malaysia's exporting of 19.9 million tons of palm oil in 2017 makes this country a world leader as an exporter of palm oil. In 2011, the country was able to generate more than 80 million tons of oil palm biomass. 30% of the 379 palm oil mills here in Malaysia utilize palm oil mill effluent (POME) by turning it to biogas [19].

4.2.3 Rice husk

Another important agricultural biomass is rice husk. This resource has a very good potential for biomass cogeneration. Biomass cogeneration refers to "generating together", this is a process where heat and EE is obtained at the same time from fuel. This type of biomass is implemented in technologies such as steam turbines, gas turbines and reciprocating engines. Currently, Malaysia has constructed its first rice husk power plant in the state of Kedah in Padang [19, 20].

4.2.4 Municipal solid wastes

For the generation of solid waste in Malaysia, the amount of mass-produced in a day range from 0.45–1.44 kg/day. This result is dependent on the economic status of the area within Malaysia. The organic waste in Malaysia contains a high amount of moisture with a bulk density of above 200 kg/m^3. The bulk density is with respect to the population growth in Malaysia, the higher the number of populations in the area the larger the bulk density of the waste. These waste are generally disposed of as landfilling which makes them ideal for being used for biomass [20] (**Figure 4**).

4.3 Biomass projects in Malaysia

4.3.1 TSH resources Berhad

The plant has huge commercial potential from its biowaste in the palm oil industry. It does this through an integrated complex system in Kunak, Tawau situated in east Malaysia, Sabah. This plant has both biomass and biogas power plants as well as being equipped with a pulp and paper plant. The plant is considered friendly to the environment as it takes into consideration the protection and preservation of the environment. This plant is responsible for the generation of electricity through the disposal of waste from oil palms. This plant is under the

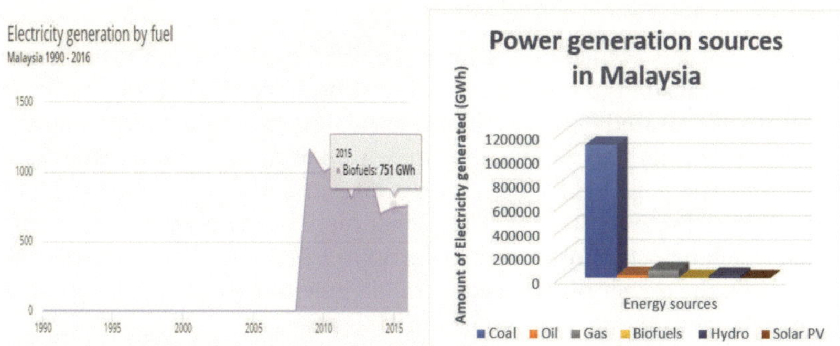

Electricity generation from 1990 to 2016 in Malaysia from biofuels.	**Electricity generation from each of the energy sources in Malaysia.**

Figure 4.
Electricity generation by fuel and Power generation sources in Malaysia.

ownership of the Kumpulan Sawit Kinabalu and has taken considerable precautions in order to create a sustainable wealth while ensuring the protection and preservation of the environment. The amount of energy production has dropped by nearly 85% since the opening of the power plant, and as a result, increasing the profit of the mills by RM1.14 million [21]. The power generation of the plant is 14 MW of completely RE from biomass cogeneration plant [21]. The plant is also the first biomass power plant connected to the main grid in Malaysia. The RE from the biomass power plant has formed an agreement with the Sabah Electricity Sdn Bhd to provide green electricity of up to 10 MW [20, 22]. TSH built up approximately 50,000 ha in a planted area across Sabah which are strategically located with associated companies. The company also has 65,000 ha worth of unplanted land bank for future development, this is to keep the company busy for many years to come. In addition to that, the company also has 3 mills in Sabah which has a 1.0 million tons of Fresh Fruit Bunches processing capacity per annum. The refining crude palm oil and kernel located at Kunak Jaya, Sabah has capacities of 2600 tons and 600 tons per day respectively [21, 22].

4.3.2 Seguntor bioenergy and Kina biopower biomass power plant

Both these power plants are in Sandakan, Sabah. Seguntor Bioenergy and Kina Biopower Power Plant are owned by HRE Seguntor Bioenergy Sdn. Bhd and HRE Kina Biopower Sdn. Bhd respectively [23]. These two power plants are implemented with a similar design. The power plants have a fuel consumption of 23.123 kg/h and a boiler capacity of 56 tons per hour at 420 degrees Celsius. Adding to that, a mechanical draught cooling tower, counterflow, water flow of around 2491 t/h. The coldwater temperature is measured at 32 degree Celsius and 42 degree Celsius for the hot water temperature. The generators are enclosed and has an in –built water – air-cooled system. A synchronous generator is also used in the system. Each power plant can provide a total of 11.5 MW of green energy from biomass energy [23, 24]. These efforts are made to ensure the provision of a stable power supply to consumers in the East Coast of Malaysia. The power plants are located strategically about 20–90 km to 15 palms oi mills. Furthermore, the power plants are located 10 km radii of SESB's substation making it ideal for grid interconnection [20, 23]. About 654,000 tonnes per year of biomass will be generated from the Nilai Tani Resources Sdn. Bhd., Monsok Palm Oil Mill Sdn. Bhd, Prolific Yield Sdn. Bhd. and Tanjung Panjang Sdn. Bhd palm oil mills. The power plants are expected to cost around RM120 million each.

4.3.3 Jana Landfill

The Jana Landfill is owned by the Jana Landfill Sdn Bhd which runs a power plant and municipal storage waste sites. It is a subsidiary of TNB Energy Services. The power plant is located in Ayer Hitam Forest Reserve in Puchong, Selangor. The Jana Landfill obtains its fuel via the decomposition of natural municipal waste from the landfill site. The power plant generates a total of 2 MW of green energy using landfill gas as a fuel source [21, 24]. The power plant has two separate sections, Jana 1 and Jana 2. Each power plant can generate up to 1 MW of energy. To date with the arrival of Jana 3, the power plants is expected to generate a combined total of 6 MW which is beneficial to about 6000 homes in Malaysia.

4.3.4 Recycle energy Sdn. Bhd

The Recycle Energy power plant is the first power in the whole of Southeast Asia to initiate Refuse Derived Fuel (RDF). RDF is the product of separating noncombustible and combustible portion from municipal solid waste. This method can help increase the recycling levels in Malaysia as well as decrease the overall waste. To add to this, this method does not require additional cost from processing, baling, wrapping and transportation logistics making it less expensive than that of the landfilling [25]. RDF can be used as a renewable fuel for any coal – fired power plant. Recycle Energy power plant is located at Kampung Pasir in Semenyih, Selangor. The plant is located about 13 miles from the main capital. This facility is able to process a total of 1100 tons of solid waste a day. These wastes are then converted into RDF which is in fluff form. The RDF in fluff form is used as biofuel to enable the production of 8 MW of electrical energy a day [24, 25]. This electricity is used to power up the RDF plant and any remaining electricity is sold for usage in the national power grid. The RDF power plant is 28 – acre wide and is handled by the Malaysian government (**Figure 5**).

4.4 Comparison between the biomass energy usages in different plants in Malaysia

When compared, the TSH Resources Berhad power plant located in Kunak, Tawau in Sabah has the highest power generation at 14 MW. The main reason for the success of the TSH Resources Berhad power plant is its immense area of coverage. Palm oil is able to be attained from the within the power plants location. Continues supplies of palm waste can be obtained and processed rapidly in the power plant. The Kumpulan Sawit Kinabalu also ensures the protection and preservation of the palm oil farms. The Seguntor and Kina power plants needs to transport its oil palm waste 20–90 km from the mills to the power plants thus having a lower efficiency rate compared to the TSH Resources power plant. Furthermore, the TSH Resources Berhad power plant uses a biomass cogeneration system enabling it to attain both electrical and heat energy from the biomass fuels (**Table 5**).

In terms of the turbines used, steam engine has a better power-to-weight ratio making them ideal for reciprocating engines. For a small size is able to generate a high amounts of power output and does not produce a lot of vibration compared to other reciprocating counterparts. The steam turbine has a higher operating efficiency and reliability compared to that of the gas turbines. Jana Landfill is the only

Figure 5.
Major biomass projects in Malaysia.

Power Plants	Power generation	Type of Turbine	Fuel	Cost	Area coverage
TSH Resources Berhad	14 MW	Steam turbines	Empty fruit bunch		5e+8 square meter (used land) and 6.5e +8 (unused land)
Seguntor Bioenergy Power Plant	11.5 MW	Steam turbines	Empty fruit bunch	RM120 million	400000 square meters
Kina Biopower Power Plant	11.5 MW	Steam turbines	Empty fruit bunch	RM120 million	400000 square meters
Jana Landfill	2 MW (From Jana 1 & 2), 6 MW (with the future addition of Jana 3)	Gas turbines	Biogas		
Recycle Energy power plant	8 MW	Steam turbines	Refuse-derived fuel		113312 square meters

Table 5.
Comparison between the biomass energy of different power plants in Malaysia.

power plant in Malaysia using gas turbines and from power generation it is the lowest compared to all the other power plants which are using steam turbines. A main issue is that, having variation in its fuel specs can lead to an enormous drop in the efficiency of the turbine. In addition to that, external power is required to ensure turbine can carry out a self-sustained operation [23, 26]. In terms of the most eco-friendly fuel source would be the source used by the Recycle Energy power plant. This power plant uses a refuse-derived fuel source making it ideal to not only for making biofuels to power up turbines inside the power plant, but also recycling resources which are not combustibles.

From the comparison table, it can be seen that in terms of practicality and usage in the whole Malaysia the TSH Resources Berhad power plant is the most ideal. It is able to generate the highest amounts of the greenest EE at 14 MW which is why it is connected to the main grid in East Malaysia.

4.5 Comparison between the usages of biomass energy in Malaysia against the World

Unlike the other countries, Malaysia is highly reliant on the usage of palm oil and coconut husk as biomass sources. Malaysia is still unable to completely utilize all resources which can lead to the production of biomass. If Malaysia can utilize converting municipal waste as a source for biomass, there is no reason to why Malaysia cannot increase their power generation from biomass. In addition to that, if Malaysia is able to produce more biomass power plants situated around palm oil fields like the TSH Resources Berhad power plant, the amount of power generated from biomass is bound to increase within the country (**Table 6**).

4.6 Section conclusion

Malaysia is still relatively new in terms of the power generation using biomass sources. From the year 2014–2016 there is a gradual increase in the power generation using biofuels in Malaysia. Thus. the potential is bright for the usage of biomass as a RESs in Malaysia.

Country	Power generation from biofuels (GWh)			Source of waste
	2014	2015	2016	
America	62357	61640	60493	Wood waste, agricultural crop, animal manure, plants and recycled waste
China	44400	52700	64700	Agricultural, forestry waste and domestic livestock (manure).
India	293926	24997	41972	Agricultural wastes.
Malaysia	701	751	760	Palm oil biomass, rice husk and municipal solid waste.

Table 6.
Comparison between the amounts of electricity generation from biomass between countries.

5. Malaysia tidal energy information

5.1 Introduction to tidal energy

Tides are created from the gravitational pull from the Earth and the moon, creating coastal tidal waters at a different time at the day. This movement of water has an enormous amount of potential energy. This energy is predictable and renewable with low operating cost. Although tidal energy is recognized as one of the promising technologies, the technology currently doesn't exist in Malaysia. There are three main types of tides phenomenon which are diurnal, semidiurnal and mixed tides [27]. Diurnal tides have one high tide every day. Semidiurnal tides have two high tides every day. Mixed tides are the combination of the characteristics of diurnal and semidiurnal tides. Tidal energy can be harnessed through different methods. A tidal barrage makes use of the tides. A barrage looks like a dam, but it's lesser height and very much bigger [28]. The other method would be the tidal stream. It works just like wind turbines, but it's placed underwater using the movement of water created by tides [27, 28]. A tidal lagoon is another method similar to the tidal barrage, but the dam is replaced by a 360-degree enclosure, creating a pool. Water will enter and exit the lagoon due to different water tides.

5.2 Feasibility of tidal energy in Malaysia

To determine the feasibility of tidal energy in Malaysia, it is required to understand the available tides in Malaysia. In Malaysia, there is no diurnal tides. The North and West of Peninsula have majority of semidiurnal tides while the area of South and East have majority of mixed tides with dominant semidiurnal. The rest of the area of Malaysia has mixed tides with dominant diurnal (**Figure 6**).

The tides differ at a different location and different times of the year. The height of water level between tides also influences the potential energy that could be harnessed. As shown in the below figure, the most potential location to harness tidal energy would be Selangor with height range between 0.4 meters to 5.3 meters when compared to other locations [28, 29].

5.3 Section conclusion

The tidal energy will always be influenced by the gravity of the moon and the sun. However, it has more advantages than wind and solar energy as it has a more predictable nature with high environmental benefits. Different locations in Malaysia have different tides and it must be considered before installing a tidal power

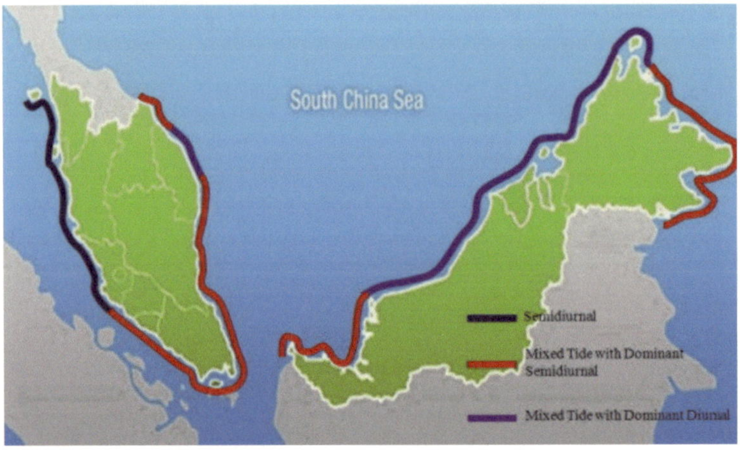

Figure 6.
Types of tides available in Malaysia [29].

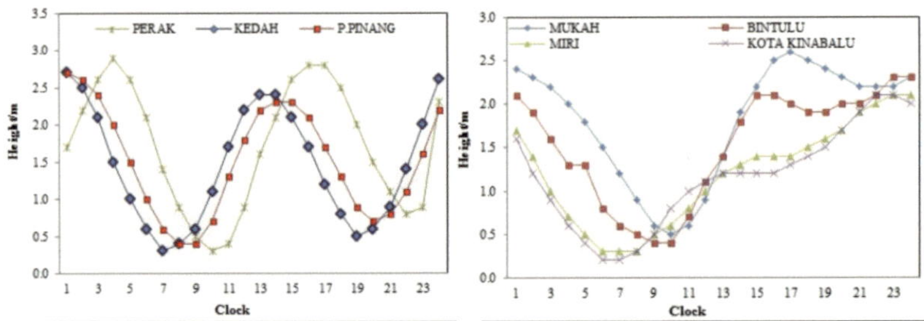

Tides in Perak, Kedah and Pulau Pinang with semidiurnal tides.

Tides from Mukah, Miri, Bintulu and Kota Kinabalu with mixed tides with dominant diurnal.

Figure 7.
Semidiurnal and Dominant semidiurnal Tides.

plant. Based on the analysis being done, Selangor has the highest potential to harness tidal energy compared to other locations in Malaysia (**Figure 7**).

6. Malaysia geothermal energy information

6.1 Introduction to geothermal energy

Geothermal energy is created by the gravitational energy of the Earth and the unstable radioactive decay of atoms [29]. Geothermal energy is mainly used to generate electricity and to provide heating. Although geothermal energy technologies have been around for over 40 years, they are still undergoing research and development. This is due to the complexity and high investments before executing geothermal projects as it includes underground exploration and requires multidisciplinary expertise [30] (**Figure 8**).

6.2 Feasibility of geothermal energy in Malaysia

Malaysia does not have technology that harnesses geothermal energy. However, there are several potential locations in Malaysia.

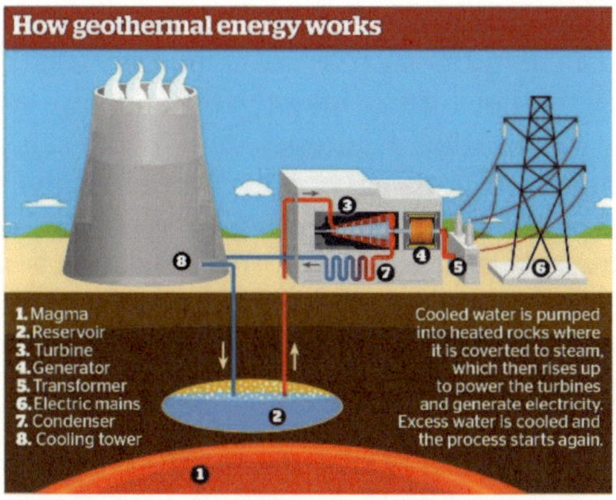

Figure 8.
How geothermal energy works.

Figure 9.
Tawau geothermal project being abandoned.

According to a study by the Deputy Natural Resources and Environment of Malaysia back in 2010, Tawau has the potential to generate up to 67 MW of electricity per day, meeting the demands of Tawau [1, 31]. The water temperatures below the selected area are near to 235 degrees which is more than enough to heat and at the same time generate electricity. This geothermal project was initiated in Tawau, Sabah, Malaysia back in 2015. However, the project site has not shown any progress and had seemed to stop operations in the third quarter of 2016. Therefore, the project's approval is being cancelled and is now currently abandoned [29, 31] (**Figure 9**).

6.3 Section conclusion

The feasibility of geothermal energy in Malaysia is inclusive of geothermal exploration and resource assessment, which requires a very high cost. The country's first geothermal power plant project failed due to the lack of preparation and great deal of discipline in executing the project. Based on the history of geothermal development, the geothermal project can easily go wrong if not all aspects are addressed adequately.

7. Conclusion

Malaysia is a developing country and being able to harness RE would be a great attribute to improve the country. The existing RE in Malaysia includes Solar energy,

Hydropower energy, and Biomass energy. Other potential REs in Malaysia could be harvested such as Tidal energy. Solar cells are commonly used to harness solar energy. The technology can be further investigated and improved to increase the efficiency of electricity generation. Hydropower in Malaysia is generating 11% of the country's electricity as Malaysia has many rivers and water bodies that could be exploited. A balanced method of implementing hydropower can be done to always ensure the ecosystem of Malaysia is not disturbed. Malaysia is still relatively new in using Biomass energy. However, the gradual increase of power generation using biofuels has increased the potential of biomass energy in Malaysia as a renewable energy source. Tidal energy has the potential to be harnessed in Malaysia as there are locations such as Selangor and Johor having tides that could generate a decent amount of electricity. Geothermal energy also has the potential in Malaysia as there are multiple hot springs. Sabah has the potential to harness geothermal energy as it originates within young volcanic area. However, more research and investment would be needed to harness geothermal energy in Malaysia. It is quite convincing that Malaysia could harness more RE as the sustainability of energy consumption is crucial in this era.

Author details

Hadi Nabipour Afrouzi[1*], Yuhani Pamodha Wimalaratna[1], Jubaer Ahmed[1], Kamyar Mehranzamir[2], San Chuin Liew[1], Chin-Leong Wooi[3] and Bazlul Mobin Siddiquea[1]

1 Faculty of Engineering, Computing and Science, Swinburne University of Technology Sarawak, Kuching, Malaysia

2 Department of Electrical and Electronic Engineering, Faculty of Science and Engineering, University of Nottingham Malaysia, Jalan Broga, Semenyih, Selangor, Malaysia

3 Centre of Excellence for Renewable Energy, School of Electrical Systems Engineering, Universiti Malaysia Perlis, Arau, Perlis, Malaysia

*Address all correspondence to: hafrouzi@swinburne.edu.my

References

[1] N. H. M Binti Tambi, H. N. Afrouzi, K. Mehranzamir, and J. Ahmed, "A review of available hybrid renewable energy systems in Malaysia," *Int. J. Power Electron. Drive Syst.*, vol. 11, no. 1, pp. 433–441, 2020, doi: 10.11591/ijpeds.v11.i1.pp433-441.

[2] R. Affandi, M. Ruddin, A. Ghani, and C. K. Gan, "A Review of Concentrating Solar Power (CSP) In Malaysian Environment," *Int. J. Eng. Adv. Technol.*, no. 2, pp. 378–382, 2013.

[3] S. A. Malik and A. R. Ayop, "Solar energy technology: Knowledge, awareness, and acceptance of B40 households in one district of Malaysia towards government initiatives," *Technol. Soc.*, vol. 63, no. May, p. 101416, 2020, doi: 10.1016/j.techsoc.2020.101416.

[4] S. Mekhilef, A. Safari, W. E. S. Mustaffa, R. Saidur, R. Omar, and M. A. A. Younis, "Solar energy in Malaysia: Current state and prospects," Renew. Sustain. Energy Rev., vol. 16, no. 1, pp. 386–396, 2012, doi: 10.1016/j.rser.2011.08.003.

[5] E. H. Miliknya, "Siaran akhbar press statement," vol. 100, p. 2662296, 2013.

[6] E. H. Miliknya, "Siaran akhbar press statement," vol. 100, no. January 2019, p. 2662296, 2013.

[7] W. S. W. Abdullah, M. Osman, M. Z. A. A. Kadir, and R. Verayiah, "The potential and status of renewable energy development in Malaysia," *Energies*, vol. 12, no. 12, 2019, doi: 10.3390/en12122437.

[8] A. Albani, M. Z. Ibrahim, and K. H. Yong, "Optimization of wind energy resource estimation in Kudat Malaysia Optimization of Wind Energy Resource Estimation in Kudat Malaysia," no. November 2018, 2013.

[9] CICERO, "'Second Opinion' on Tadau Energy's Green Sukuk Framework," no. July, 2017.

[10] T. S. Khoon, "A Renewable Power Source: Visit to Amcorp Gemas Solar Power Plant," no. September, p. 2015, 2015, [Online]. Available: http://dspace.unimap.edu.my/dspace/bitstream/123456789/40810/1/A renewable power source- visit to amcorp gemas solar power plant.pdf.

[11] G. Masson and M. Brunisholz, "2015 Snapshot of global photovoltaic markets," *Iea Pvps T1–292016*, pp. 1–19, 2016, [Online]. Available: http://www.iea-pvps.org/fileadmin/dam/public/report/PICS/IEA-PVPS_-__A_Snapshot_of_Global_PV_-_1992-2015_-_Final_2_02.pdf.

[12] A. Jäger-Waldau, "Snapshot of photovoltaics-February 2020," *Energies*, vol. 13, no. 4, 2020, doi: 10.3390/en13040930.

[13] M. Hossain *et al.*, "A state-of-the-art review of hydropower in Malaysia as renewable energy: Current status and future prospects," *Energy Strateg. Rev.*, vol. 22, no. November, pp. 426–437, 2018, doi: 10.1016/j.esr.2018.11.001.

[14] Malaysia Energy Commission, "National Energy Balance 2016," *Energy Comm.*, pp. 1–114, 2018, [Online]. Available: www.st.gov.my.

[15] A. Kadier, M. S. Kalil, M. Pudukudy, H. A. Hasan, A. Mohamed, and A. A. Hamid, "Pico hydropower (PHP) development in Malaysia: Potential, present status, barriers and future perspectives," *Renew. Sustain. Energy Rev.*, vol. 81, no. June 2017, pp. 2796–2805, 2018, doi: 10.1016/j.rser.2017.06.084.

[16] K. H. D. Tang, "Hydroelectric dams and power demand in Malaysia: A

planning perspective," *J. Clean. Prod.*, vol. 252, no. December, p. 119795, 2020, doi: 10.1016/j.jclepro.2019.119795.

[17] M. Kadir, D. Ghazali, and T. A. Musa, "Pergau Reservoir Information System (PRIS) For Mapping and Sedimentaion Studies : A Study on the Development of the Reservoir Database," Asian Conf. Remote Sens., pp. 1–7, 1997.

[18] S. Tang, J. Chen, P. Sun, Y. Li, P. Yu, and E. Chen, "Current and future hydropower development in Southeast Asia countries (Malaysia, Indonesia, Thailand and Myanmar)," *Energy Policy*, vol. 129, no. September 2018, pp. 239–249, 2019, doi: 10.1016/j.enpol.2019.02.036.

[19] S. M. Shafie, T. M. I. Mahlia, H. H. Masjuki, and A. Ahmad-Yazid, "A review on electricity generation based on biomass residue in Malaysia," Renew. Sustain. Energy Rev., vol. 16, no. 8, pp. 5879–5889, 2012, doi: 10.1016/j.rser.2012.06.031.

[20] M. D. M. Samsudin and M. M. Don, "Municipal solid waste management in Malaysia: Current practices, challenges and prospect," *J. Teknol. Sciences Eng.*, vol. 62, no. 1, pp. 95–101, 2013, doi: 10.11113/jt.v62.1293.

[21] Y. H. Chan *et al.*, "An overview of biomass thermochemical conversion technologies in Malaysia," Sci. Total Environ., vol. 680, pp. 105–123, 2019, doi: 10.1016/j.scitotenv.2019.04.211.

[22] S. H. Shuit, K. T. Tan, K. T. Lee, and A. H. Kamaruddin, "Oil palm biomass as a sustainable energy source: A Malaysian case study," Energy, vol. 34, no. 9, pp. 1225–1235, 2009, doi: 10.1016/j.energy.2009.05.008.

[23] Z. Haryati, S. K. Loh, S. H. Kong, and R. T. Bachmann, "Pilot scale biochar production from palm kernel shell (PKS) in a fixed bed allothermal reactor," *J. Oil Palm Res.*, vol. 30, no. 3, pp. 485–494, 2018, doi: 10.21894/jopr.2018.0043.

[24] M. S. Umar, P. Jennings, and T. Urmee, "Generating renewable energy from oil palm biomass in Malaysia: The Feed-in Tariff policy framework," Biomass and Bioenergy, vol. 62, pp. 37–46, 2014, doi: 10.1016/j.biombioe.2014.01.020.

[25] C. Nobre, O. Alves, L. Durão, A. Şen, C. Vilarinho, and M. Gonçalves, "Characterization of hydrochar and process water from the hydrothermal carbonization of Refuse Derived Fuel," Waste Manag., vol. 120, pp. 303–313, 2021, doi: 10.1016/j.wasman.2020.11.040.

[26] J. Zueco, D. López-Asensio, F. J. Fernández, and L. M. López-González, "Exergy analysis of a steam-turbine power plant using thermocombustion," *Appl. Therm. Eng.*, vol. 180, no. July, p. 115812, 2020, doi: 10.1016/j.applthermaleng.2020.115812.

[27] N. A. Mohd Yusoff, N. L. Ramli, and M. R. Mohamed, "Investigation of the potential harnessing tidal energy in Malaysia," ARPN J. Eng. Appl. Sci., vol. 10, no. 21, pp. 9835–9841, 2015.

[28] A. L. Maulud and H. Saidi, "The Malaysian Fifth Fuel Policy: Re-strategising the Malaysian Renewable Energy Initiatives," Energy Policy, vol. 48, pp. 88–92, 2012, doi: 10.1016/j.enpol.2012.06.023.

[29] S. E. Ben Elghali, R. Balme, K. Le Saux, M. El Hachemi Benbouzid, J. F. Charpentier, and F. Hauville, "A simulation model for the evaluation of the electrical power potential harnessed by a marine current turbine," IEEE J. Ocean. Eng., vol. 32, no. 4, pp. 786–797, 2007, doi: 10.1109/JOE.2007.906381.

[30] J. W. Lund and A. N. Toth, "Direct utilization of geothermal energy 2020

worldwide review," *Geothermics*, vol. 90, no. July 2020, p. 101915, 2021, doi: 10.1016/j.geothermics.2020.101915.

[31] M. B. Farriz, A. N. Azmi, N. A. M. Said, A. Ahmad, and K. A. Baharin, "A study on the wind as a potential of renewable energy sources in Malaysia," *ECTI-CON 2010–2010 ECTI Int. Conf. Electr. Eng. Comput. Telecommun. Inf. Technol.*, pp. 651–655, 2010.

Chapter 3

Itaipu Technology Park: An Eco-Innovative Niche for Renewable Energies

Andriele De Pra Carvalho and Sieglinde Kindl da Cunha

Abstract

Technology parks are considered innovative environments for the development of new technologies, in a dynamic that can be explained by the micro level of multilevel analysis of the theory of sociotechnical transition, by contributing to explain the actors responsible for the process of development and dissemination of technology. The cases analyzed were composed of companies and eco-innovative projects in the renewable energy area of the Itaipu Technological Park. The methodology was based on a study of multiple cases of qualitative nature, supported by content analysis and triangulation of information. The main results showed that the ITP, with the support of its maintainer Itaipu Binacional, created an environment conducive to the development of eco-innovative companies in renewable energies, stimulating the exchange of knowledge between companies and favoring partnerships with national and international companies for technological development.

Keywords: Itaipu Technology Park, renewable energy, micro level, multilevel analysis, eco-innovation

1. Introduction

The environmental problems arising from uncontrolled growth have broadened the interest in studies that combine innovation with sustainability. Innovative technological trajectories began to seek efficient ways of environmental preservation, combined with organizational competitive advantages [1]. In this sense, linking innovation with environmental sustainability, as in the so-called eco-innovation, a concept addressed in the studies by CARRILLO-HERMOSILLA, J.; GONZALEZ, P. R.; KONNOLA, T [2], contributes to understanding the relationship between society, the economy and the environment.

This transition to sustainability or eco-innovation, according to GEELS [3] can be studied by multilevel analysis that highlights that there are different levels by which eco-innovation carries out its trajectory. The micro level, which is the level of niches and actors responsible for stimulating and beginning the eco-innovative process, the meso level, which is the dominant regime and the macro level, which accounts for changes at the global and macro level.

Because multilevel analysis, at its micro level, involves technological niches that act as a space for relationships between multiactors, being conducive to radical innovations and the interactive learning process, this article directs its focus to this level.

The micro level also contributes to the construction of social networks to support innovation, in a strategic management [4–6]. Thus, the socio-technical regime and the environment in which organizations are inserted interfere in the development of ecoinnovations at the level of technological niches. In this perspective, the context of a Technological Park stood out for the realization of this study for combining the sociotechnical regime and the development of ecoinnovations, as well as for housing companies, niches and environments with different characteristics [7].

In this context, the Itaipu Technological Park - PTI, based on the sustainability report of its maintainer, Itaipu Binacional, has become an environment conducive to the development of this study, because it is a scientific and technological pole based on a model of support for sustainable technologies and practices, which has as its theme of interest in its Strategic Planning the energy sector, an essential technology for a country and relevant for the development of renewable energy sources that consider the premises of the reduction of the environmental impact.

The discussion then focused on the transition to sustainability favored by energy eco-innovation niches developed in the Itaipu Technological Park. Thus, in this article we bring the theoretical lens of the micro level of the sociotechnical system [8, 9] in the discussion on the development of renewable energies in the Itaipu Technological Park (PTI) since this approach considers that the technological transition begins its process at the micro level. Thus, the following question arises: How does the micro level of THEPT contribute to the development of eco-innovations in renewable energies?

2. Methodology

The sociotechnical approach is a medium-range theory that crosses different ontologies, because it is a multidimensional phenomenon that can be studied from various angles and by different disciplines [9]. There are different epistemological assumptions for its study, although the theory is based on particular intersections between the theory of evolution, interpretivism and constructivism and structuralism, besides combining the evolutionary view, the dynamics of systems, sociological view of innovation and the understanding that multilevel narrative always assumes the historical dimension.

The direction of this study followed the qualitative method as a research strategy, through the analysis of social aspects, which involve interpretation and that are researched in its natural environment, that is, in loco.

The first phase of the research used the case study approach [10], corresponded to the exploration phase of the empirical field to seek relevant answers to the research question. For the choice of objects of analysis, according to STAKE [10], because it is a case study and a qualitative approach, we opted for extreme cases of relevance. The PTI, because it is a park created by the Itaipu Hydroelectric Power Plant, is highlighted in the creation and dissemination of knowledge and for having projects in the area of energy, which are part of the interest and planning of the PTI, besides being an important actor in this segment for the country. Thus, the study will consider the following cases of the Park in the energy area: (i) a company incubated in pti, AP energia, which acts in improving the energy efficiency of generators, the only company incubated in this energy sector; (ii) the graduate company Esco Iguassu, which operates in the area of Energy Management, is located in the city of Foz do Iguaçu and continues to partner with Itaipu Binacional and itaipu technological park; (iii) the project that is in process to become a company, Cibiogás, which acts as an international center of renewable energy, through

Biogas; and (IV) the Hydrogen project, which seeks in this chemical component, ways to contribute to the energy matrix and energy storage.

The investigation occurred inductively, using the techniques of interview, observation and documentary analysis [10, 11], in addition to informal conversations with employees and project participants in the energy area within the PTI: AP Energia, Esco Iguassu, Projeto Hydrogen and Cibiogás.

The observation was developed in a period of forty hours in all activities carried out in the Park. The documentary research involved documents made available by the park and also information on the websites of PTI, FPTI and Itaipu Binacional.

The interviews referenced in this research were conducted with four PTI managers and the four managers of the energy niches of the PTI: AP energia, Cibiogás, Esco Iguassu and Projeto Hydrogen.

The second phase of the research was dedicated to data analysis, with the help of atlas IT software, being characterized as descriptive and analytical, for reporting practices, facts, evidence and statements in order to consolidate information that allowed the analysis of the micro level of the PTI based on the Geels Multilevel Perspective [4, 8, 12], consistent with the sociotechnical transition approach.

3. Development (results and discussions)

The Itaipu Technological Park was created in 2003 by Itaipu Binacional, in the city of Foz do Iguaçu, in western Paraná, bordering Paraguay and Argentina [13, 14].

The former space that served as accommodation for itaipu binacional builders was transformed to accommodate the activities of the PTI. The physical infrastructure adapted to the reality of the PTI covers classrooms, language laboratories with computers, academic computer labs, videoconference rooms, teleroom, library, study and support rooms, sports courts, academic laboratories, auditorium, three event spaces, support room, cyber room, cinetheater, incubator and business condominium [13].

In this environment, this study was carried out that follows the context of Geels' sociotechnical theory (2014), in which the development of an eco-innovation in renewable energy begins at the micro level, which is the level of niches, in which the relationship between the actors that will stimulate the development of new technology occurs and to house companies, niches and environments with different characteristics [7].

In technology parks, different interactions between actors at the micro levels can be perceived, which are fundamental to understand the dynamics of eco-innovation in renewable energy. The actors of the technological parks that make up the micro level can be represented by technology-based companies, science and technology entity, government, governing bodies, funding agency, developers, universities and research centers, information technology centers, consultancies, stakeholders, entities and local business community [7]. In practice, it is noted that these actors compose the infrastructure of the ITP, presenting a certain complexity in their relationships, some of which can be observed in **Figure 1**, which contemplate the actors of the ITP and their relationships.

These actors act as stimulators for the development of renewable energy niches, which are commented in each case analyzed in this study.

3.1 The development of renewable energies in the PTI niche

The different actors of the Park, these act in conjunction with the cases analyzed, which are the ecoinnovations in Renewable Energies. AP Energia is the

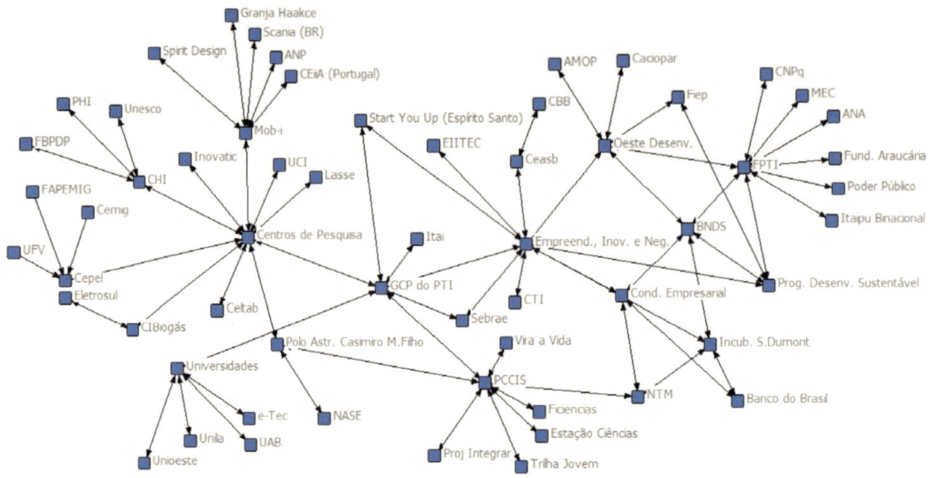

Figure 1.
Multiactors of the Itaipu Technological Park. Source: Prepared by the authors.

first company analyzed, whose description was based on the interview conducted with its general manager, having been complemented by data from documentary research, observation and field notes.

The first analysis sought to group the codes that related to the category "AP Energia", through the Atlas.TI software, to interconnect the main relevant points of the company and its structure, as shown in **Figure 2**. Founded in 2014, AP Energia's main purpose is to target segments of electricity generation from renewable sources. It serves two large segments consisting of distributed generation and medical records of industrial electrical installations. From these medical records, the company performs financial feasibility analysis, electrical studies, NR-10 medical records, quality analysis and electrical projects. However, the activities of medical records of electrical installations are complementary works to the innovative technology of the company, which represents distributed generation, which works with complete consulting for the connection of electric power generators in distribution networks. In other words, it aims to develop technologies and services for the connection of renewable energy sources, from the producer (individual) to the concessionaire.

This aims at the constant improvement of its technology, seeking to be a reference in studies of electricity generation from renewable sources or not and the provision of continuous electricity services. The company's location is in the Business Condominium within the Itaipu Technology Park. The room is small, because the company does not require large space, because it works only with software in the area. In addition, it has no employees, since much of its technology began through a graduate project, at the doctoral level, which continues to be running, as the company's manager attests.

The company was born from the academic context and still maintains this bond today. The Park acts as a connection of the education acquired for practical application, a fact that becomes clear in the words of the manager, when it intensifies that the Park allowed to put into practice its studies and be an entrepreneur, mainly in projects aimed at improving the environment, which align the market with renewable energy sources.

All these developments and activities aimed at sustainable innovation depend on actors and existing relationships at the micro level, which covers the beginning and definition of the eco-innovation process. This category of analysis is composed of the codes shown in **Figure 3**.

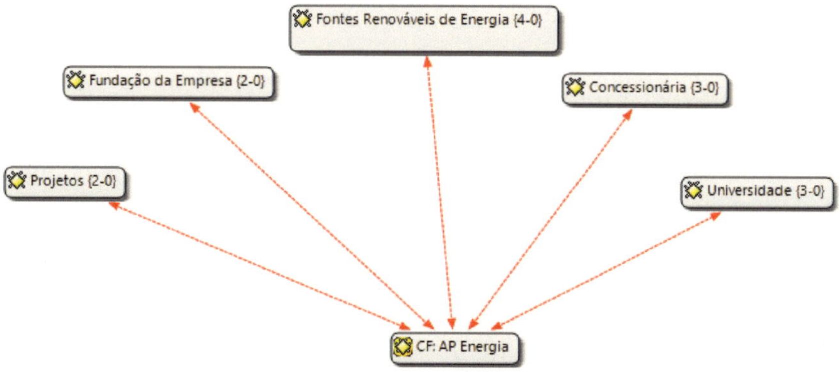

Figure 2.
Category or family "AP energy". Source: Own elaboration.

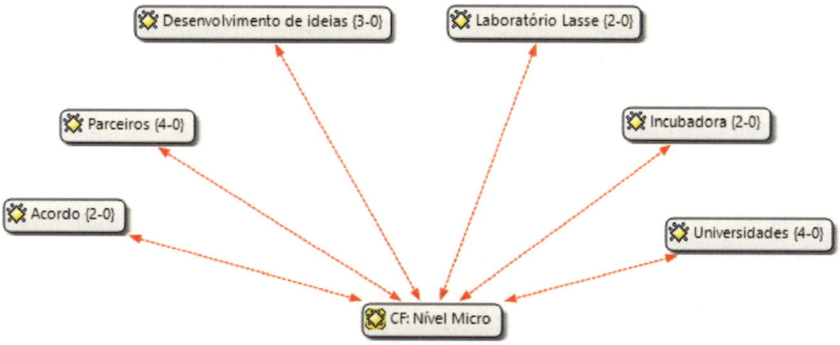

Figure 3.
Category or family "micro level". Source: Own elaboration.

According to the manager of AP Energia, the actors of the Park have contributed to the development of eco-innovation. This is because sustainable innovation usually follows a process started from an idea, but which is still uncertain in its future development, so the contribution of the park's actors is fundamental. According to the clarification of the manager of AP Energia, the Santos Dumont Incubator, one of the actors of the Park, was fundamental for its development, through subsidies, consulting, training, training and contact networks. The subsidies were passed through space and physical structure to the company. The consultancies, training and training helped the management of the company, because the incubator aims to strengthen its teams for the competitive market, according to one of the incubator managers.

At this point, it remains evident that the park managers' view of directing the company to the market is recognized by the ap energia manager as beneficial. There is a relationship of respect and hierarchy very clear, because the PTI has great national and international visibility and, as indicated by the manager of AP Energia, by loading the name of the Park along with the name of his company, facilitates the opening of new doors in the market, mainly for partnerships. The importance of the partnerships was highlighted by the manager: "As my technology depends on the investment of concessionaires, the name of PTI has contributed to open doors, agreements and negotiations with partners." It is noted how the influence of actors at the micro level is important for the dissemination of information and relationships necessary for innovation to happen through partnerships. Thus, another partner and actor considered important by the manager of AP Energy was the Lasse laboratory, located in PTI and fundamental support for the development of the company's technology.

As the company's innovative technology is in the process of evolution, it is observed how PTI instigates research and innovation. Because it is an academic environment, where students and teachers circulate among several researchers and partners of the park of national and international reputation, in a place of interaction and learning.

The second company analyzed, Esco Iguassu, is already graduated from the Santos Dumont Incubator of PTI and operates in the energy area. The first analysis carried out with the help of atlas ti grouped the codes directed to the category "Esco Iguassu", which addresses the main relevant points highlighted during the interview, which are highlighted in **Figure 4**.

Esco Iguassu started its incubation process at PTI in 2008 and, at the end of 2012, received its graduation, determining the end of the incubation period and the beginning of autonomy for the company. However, it remained installed in the premises of the Itaipu Technological Park. This fact demonstrates how interesting it is for the park, mainly because it also acts as a partner of the Study Center on Biogas, one of the broader projects of the PTI, as well as, because its activity is directed to the energy area, one of the three clusters that define the direction of the Santos Dumont Incubator, as indicated by its managers.

As already explained, Esco Iguassu operates in the energy area and aims to promote energy efficiency in its concession area. Therefore, it benefits from some national laws and the provision of government resources for this purpose through the Energy Efficiency Actdoes not. 10,295, 2011, which provides for the National Policy for Conservation and Rational Use of Energy.

According to the company's manager, the projects carried out meet the current theme with a focus on reducing energy consumption. The energy theme emerged as the main idea for the opening of the company, because its manager had been working in this area since he was graduating, within the PTI. Currently, he continues with research in the area, such as graduate student, master's level. As in the case of AP Energia, there is a strong interaction between the university and the Technological Park driving entrepreneurship.

The activity developed by Esco Iguassu demonstrates its understanding of the eco-innovative practices witnessed in the ITP, which are manifested in the actions and languages built. All these practices are influenced by the actors and dimensions of the perspective of the micro level, which can be configured as small market niches, in which the habits and routines present between individuals and the environment is conducive to new experiences and innovations. The category of the micro level family is expressed in **Figure 5**.

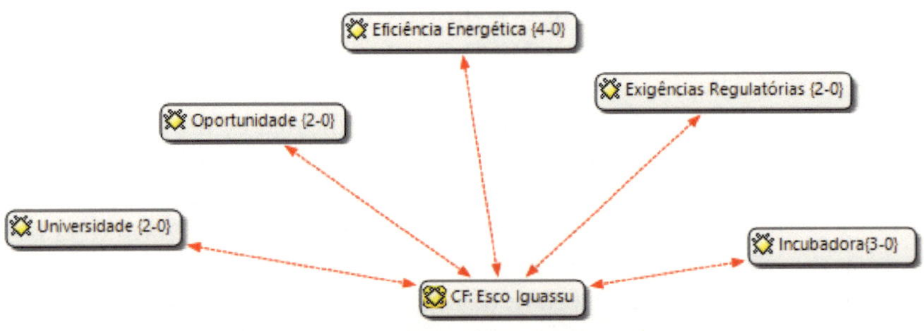

Figure 4.
Category or family "Esco Iguassu". Source: Own elaboration.

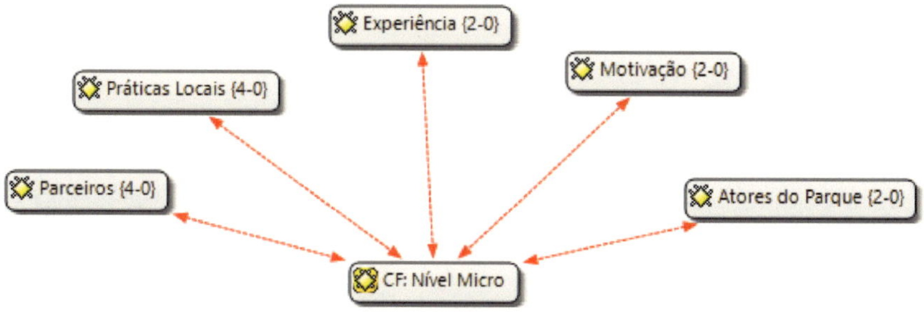

Figure 5.
Category or family "micro level". Source: Own elaboration.

The environment of the Itaipu Technology Park is repeatedly cited by the manager of Esco Iguassu for having been fundamental to the development of his idea when he began graduation. This relationship occurs through universities, students, partners and other actors of the Park.

Among the benefits of the environment, pti also assisted the company in relation to traditional financial support, passed on to all companies in the Park, such as: rent, maintenance expenses and overheads, because according to its manager, the company never had financing or other source of exclusive resources. The environment, therefore, stimulates through its practices and not only by the transfer of financial resources.

The third case studied is the International Center for Biogas Studies- Cibiogas. In view of its scope, Cibiogás was also a case analyzed in the study conducted by MENDONÇA [15], which focused on the factors of the multilevel relationship in the process of sociotechnical transition to eco-innovation in Itaipu Binational programs. Study that will also serve as a reference for this description.

Thus, through the collected data, the first category of Analysis formulated was "CIBIOGÁS", which grouped the codes in **Figure 6**, highlighting the notes on the characterization of the company.

The International Center for Biogas Studies, Cibiogás, is an organization specialized in consulting, knowledge sharing and labor analysis in renewable energies, with emphasis on biogas. In addition, the entity promotes the development of projects and public policies related to the theme, with the aim of encouraging the generation of biogas in a sustainable and renewable way [16].

Cibiogás began as a prominent itaipu project, when the plant included in its organization chart the Renewable Energy Advisory. The project was based on the partnership between Onudi, Eletrobras and Itaipu, which enabled the creation of the Renewable Energy Observatory involving all of Latin America and the Caribbean, boosting the creation of the Biogas Laboratory and, later, the Center for Biogas Studies, based on a specific methodology of the University of Tera in Vienna, Austria, and structured by standards of organization of International Technology Centers [15].

At the World Energy Conference in 2011, Cibiogás was presented by itaipu's director, the director general of Onudi, through a letter, the reasons for its application, which was well received. Subsequently, the final presentation occurred at the Center in Rio + 20 through a protocol of intention of the development of Biogas studies, which was signed by the authorities present [15].

Cibiogás was legally characterized as an international, non-governmental and non-profit organization, a specific purpose company, with autonomy to acquire movable property, real estate and participation. The Biogas Laboratory was installed next to the Itaipu Technological Park in 2011, due to the fact that

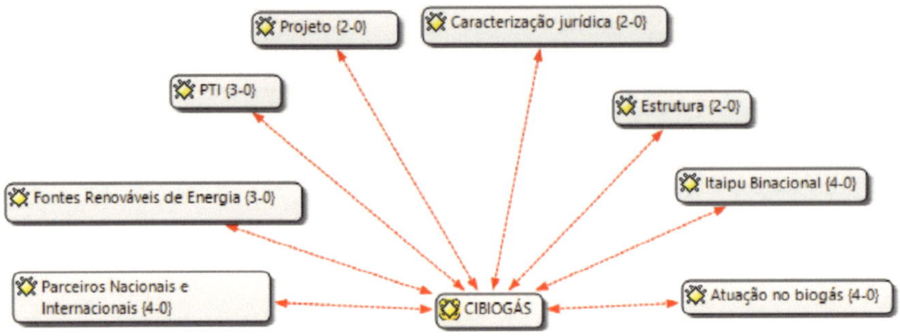

Figure 6.
Category or family "CIBIOGÁS". Source: Own elaboration.

the Park began to sign the agreements outlined by Itaipu Binacional when it began to expand its operations, through events, seminars and tests for the use of Biomethane in partnership with Scania do Brasil [16]. According to its director, the Center has grown a lot, won the ISO 9001:2008 (Quality Management System) certificate and has several partners, as well as, it is moving towards being effected as a company.

Cibiogás is a private non-economic association with seventeen associates and partners who are working together, in synergy, to consolidate Cibiogas in Paraná and Brazil, among them: Itaipu Binacional, PTI, Eletrobrás, Onudi, Onu, Eletrobrás/Cepel, Toledo Prefecture, Cooperativa Lar, Sebrae, FIEP, Seabe, Iapar, Federation of Agriculture of the State of Paraná (FAEP), Embrapa, Sebrae and United Nations Food and Agriculture Organization (FAO).

Its head office is in the ITP and has the following structure: a biogas laboratory, eleven national demonstration units and small and medium rural properties in the region and an international demonstration unit in Uruguay [16].

All this relationship between actors composes the variables of the category "Micro Level of Cibiogas", reported in **Figure 7** below.

The micro level of Cibiogas is characterized by the main evidence extracted from data collection. This level identifies the main partners and actors of the Center. Among this network, Itaipu Binacional and the Itaipu Technological Park received great importance, because they are the creators and main supporters of the project. In addition to these, other actors and partners were also classified as fundamental to the Center, which have already been commented at the beginning of this analysis and identified in the Cibiogas Implementation Project [17]. These actors interact with each other and between existing technologies and opportunities, initiating a network of cooperation.

It is evident in his speeches the importance that the director of Cibiogás and his collaborators attribute to the partnerships and actors involved in the exchange of knowledge, research and experiments. An example of this partnership is universities, which state that they contribute to the nationalization of technology with studies aimed at improving the process and resolving inconsistencies with the support of academic research [15].

In addition to the participation of all these actors, the micro level also highlights the participation of the population as a collaborator for the development of innovation, especially after recognizing the importance of technology for the region and for local cooperatives.

The fourth case analyzed, the Hydrogen project, is part of the Itaipu Technological Park highlighted due to its eco-innovative potential. Thus, based on data collection, the first analysis performed with the help of Atlas.Ti, grouped the codes of the category "Hydrogen Project", as shown in **Figure 8**.

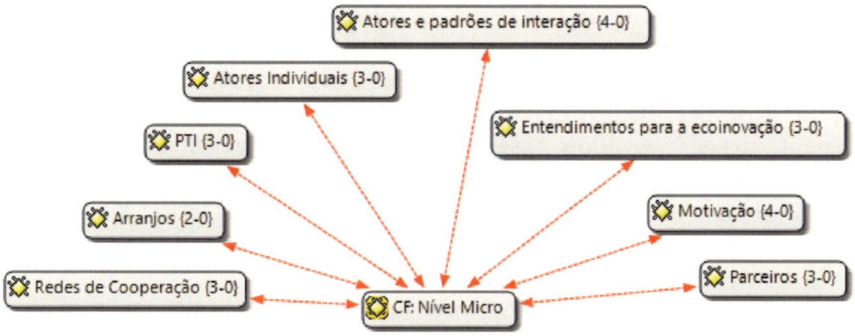

Figure 7.
Category or family "micro level". Source: Own elaboration.

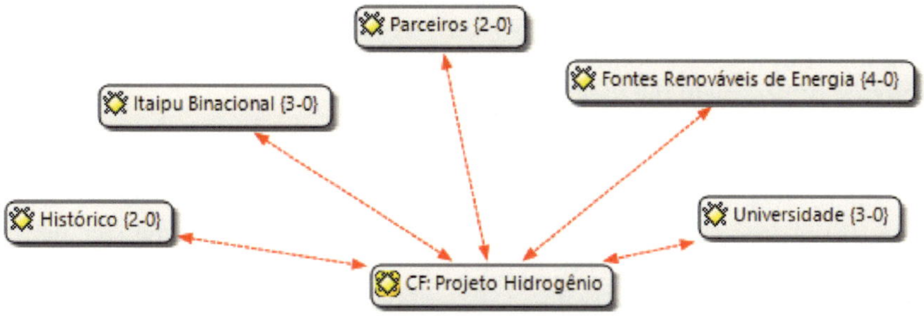

Figure 8.
Category or family "hydrogen project". Source: Own elaboration.

The project began in 2010 with specialists from Itaipu Binacional, itaipu technological park and Eletrobras, with the purpose of implementing an experimental hydrogen production plant and a research group in the area so that, from hydrogen capsules, it stores energy. That is, to analyze the viability of hydrogen production from renewable energy sources and store it in cylinders, in the form of gas, to be used in fuel cell and produce electricity. This, in turn, would be used to supply homes, industries, electric vehicles or even as a backup system, for its capacity to store energy, this being the main differential.

The stimulus to the project was given through the master's study of engineer Antônio Carlos Fonseca, manager of the Department of Electronic and Electromechanical Engineering of itaipu's Superintendence of Engineering, developed with Unicamp, at the National Reference Center for Hydrogen Energy. It was developed in line with Itaipu's strategic planning and its objective: renewable energies. According to the project director, its direction was bureaucratic, but the financial support of Itaipu and Eletrobras was decisive for the start of the plant's operation in 2011 and in 2014 it began to produce hydrogen [17].

Thus, as described by its director, the hydrogen project being consolidated will supply a national demand, as it will enable the storage of energy generated from renewable sources, such as hydroelectric plants, solar and wind power. According to Itaipu [17], the main expected benefits with the project are: to allow the evaluation of hydrogen production from hydroelectric power; provide infrastructure for hydrogen technology research; evaluate the reduction of water and energy waste in hydroelectric plants and evaluate the potential to reduce environmental impact; provide, in the technological field, research, development and innovation.

Another aspect highlighted by the project director, concerns the use of hydrogen in the form of fuel cell for engine operation, this made him seek the partnership of academics from universities installed in the PTI to develop studies in this area.

In this analysis, it is noticeable how the project is still being modeled and structured, with the contribution of several actors that make up the micro level, one of the categories of analysis, presented in **Figure 9**, with its codes grouped.

The micro level of the hydrogen project is related to the development and implementation of innovation, based on social processes, existing practices and the interaction between the multiple actors, fundamental to the new technology. An example of these individual actors is the Itaipu Hydroelectric Power Plant, which, through its strategic planning, seeks ways to encourage renewable energy sources, such as a demand of its own.

Thus, when the plant's engineer, Antônio, presented his project focused on the hydrogen plant, Itaipu began the process so that it could be put into practice. The lack of resources weighed as one of the obstacles to the beginning of the project, however, Eletrobras, another actor, stepped in and financed a part of the project, which may finally become a reality [17].

For the project director, partners are key to consolidating. Through the partners, you can direct new professionals, new financing, new opportunities and actions that further disseminem the idea of the project.

Innovation developed at the niche level is focused on learning, collective practice and the creation of cooperation networks.

The micro level, analyzed in the four cases of this research, allowed the understanding of the processes that initiate the strategy that leads to the management of the transition to a new regime [18]. In the cases analyzed, it was evident how they protect the spaces for the development of eco-innovations, which makes it a learning community allied to many actors directed to entrepreneurship, innovation and sustainability.

The technological niches in renewable energies present in the ITP demonstrate that the transformations that occur there are engaged by the collective perception of opportunities, brought through the relationship between meso and macro levels [19].

3.2 Relationship between PTI actors-partners for sustainable energy development

The technological niche discussed in this study, in a private environment, in this specific case, the Itaipu Technological Park, proved to be conducive to the development of partnerships between several actors, which were paramount for the exchange of knowledge and the development of ecoinnovations [6, 20]. These partnerships highlighted in the four cases analyzed, evidenced how the actors of the Park cooperate and complement each other in the activities to achieve the objectives of the ITP. In these relationships, in view of the observations made, the manifestations

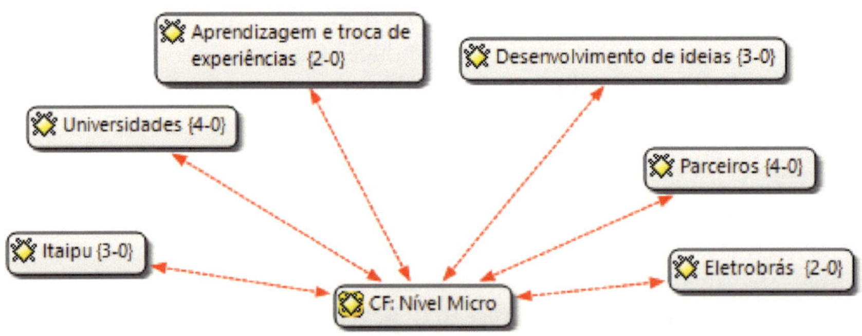

Figure 9.
Category or family "micro level". Source: Own elaboration.

of the managers of the ITP, the Incubator and the companies, no competition was identified between them, only partnership and mutual collaboration.

Cooperation and mutual growth are also identified in the reports of the managers of the Santos Dumont Incubator, when they pass on that the understanding of incubated companies during the selection process is that the park environment seeks cooperation and mutual growth among the actors, mainly allied to the objectives of the PTI. In the training and consultancies passed on to companies, the exchange of experiences and partnerships are also encouraged, one of the dimensions of the micro level, according to GEELS [3] and DOLATA [21]. It should be noted that, during the descriptions of the cases, it is clear how the Itaipu Technological Park attracts many partners, something beneficial for both companies and projects.

The cases Esco Iguassu and AP Energia state that they have other formal partners that have contributed to eco-innovation, also through the PTI. These two cases recognize that the name PTI and Itaipu opened many doors and served as a rapprochement with many partners. While the Hydrogen and Cibiogas projects, because they were created by Itaipu, had many partners before they even left the paper. This fact demonstrates the incentive of the plant in the dissemination of new sources of the energy matrix, influenced by pressures of the landscape level and even the level of the regime [3, 22].

It is common to have several projects with two or more actors working in partnership and in collaboration in the search for innovations, mainly in the topics of interest to the Park: water, energy and tourism. There is a culture of sharing between projects that comprises the common use of the Park's infrastructure, which causes them to gradually form collaborative and learning networks [19].

The partners interconnected with eco-innovative technologies in the energy area are composed of national and international actors. The incubated companies AP Energia and Esco Iguassu have only a partnership with national actors, who, according to their managers' report, were very important for the development of the technology. These partners were conquered mainly through the influence of the names PTI and Itaipu, which also act as partners [3, 12]. At this point, we understand the care that the PTI has for the selection of incubated companies, because it takes into account that it will follow related to its name and the name of Itaipu.

The Cibiogás and Hydrogen projects, developed and stimulated by PTI and Itaipu Binacional, in addition to national actors, also have the partnership of international actors, which demonstrates their primacy for the sociotechnical transition process. The main partners of these projects were international, for contributing to the supply of components and, in particular, to the exchange of knowledge with similar technologies developed outside the country.

When analyzing these actors, it is noted that there are related multiactors in all projects, that is, a cooperation network. In this premise, we highlight the articulation of these actors to establish a nucleus of partnerships, which are important in the development for the process of sociotechnical transition to the new regime, because, according to GEELS [3], at the micro level begins the process of dissemination of new technology. From this perspective, GEELS and KEMP [20] also emphasize that the role of multiactors, partnership networks and collaboration networks are fundamental for the dissemination of new technology, in this case, of the new energy matrix. **Figure 10** represents this network of partnerships.

Among the actors identified, the Itaipu Hydroelectric Power Plant has the greatest influence and prominence in the development of the Park, a fact proven during the collection of data from the park managers and companies and projects of the analysis cases. This reinforces the understanding of itaipu power plant's intention to expand its potential to new sources of Energy Matrix, which may shape new configurations for a new sociotechnical regime [3, 4].

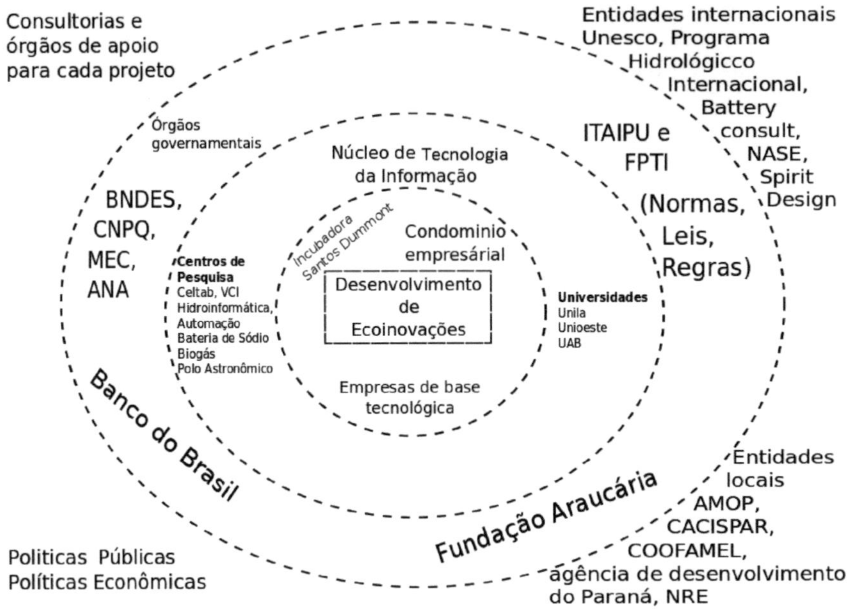

Figure 10.
Pti partner multiactors. Source: Own elaboration.

4. Final considerations

In these analyses it is evident how the environment of the Park is conducive to the development of partnerships between the various actors, who cooperate with each other and complement each other, becoming paramount for the exchange of knowledge and development of ecoinnovations.

It is common to have several actors acting through partnership or collaboration in the topics of interest to the Park, among them, energy. The actors common to all cases analyzed in this study are: Itaipu Binacional, PTI, Eletrobras, Sebrae, CNPq, PTI universities, PTI academics, ANEEL and the federal government. It is noted how these actors common to all projects are national. International actors work in partnership only with specific projects such as the Hydrogen Project, the University of Ukraine and Cibiogás, whose partners are the UN, Onudi, the University of Vienna, FAO and the University of Ukraine.

As is expected Itaipu Binacional is the actor with the greatest influence of the Park, for attracting national and international partners for its projects, which in partnership with internal actors, develop ecoinnovations. At this point, the PTI infrastructure is also a highlight to attract partners and stimulate a culture of sharing, collaboration and learning, which reinforces the potential of PTI for the development of new energy sources. This potential is also identified in the internal processes of management of energy niches, which involves the relationships of these actors to develop management strategies for the sociotechnical transition to sustainability. In the cases analyzed, the involvement of the actors were important for the investment of intellectual and financial efforts in the development of technologies, in addition to the exchange of knowledge and experiences, which were paramount for the development of the energy niche.

Author details

Andriele De Pra Carvalho[1*] and Sieglinde Kindl da Cunha[2]

1 The Federal University of Technology – Paraná, Francisco Beltrão, Brazil

2 Positive University, Curitiba, Brazil

*Address all correspondence to: andridpc@gmail.com

References

[1] CORAZZA, R.; FRACALANZA, P. S. Caminhos do pensamento neo-schumpeteriano: para além das analogias biológicas. Nova Economia. Belo Horizonte. v. 14, n. 2, p. 127-155, 2004.

[2] CARRILLO-HERMOSILLA, J.; GONZALEZ, P. R.; KONNOLA, T. Eco-innovation: when sustainability and competitiveness shake hands. [S.l.]: Palgrave Macmillan, 2009.

[3] GEELS F. A socio-technical analysis of low-carbon transitions: introducing the ultilevel perspective into transport studies. Journal of Transport Geography, v. 24, p. 471-482, 2012.

[4] Geels F. Co-evolutionary and multi-level dynamics in transitions: the transformation of aviation systems and the shift from propeller to turbojet (1930-1970). Technovation, v. 26, p. 999-1016, 2006.

[5] GENUS, A.; COLES, A. M. Rethinking the multi-level perspective of technological transitions. Research Policy, v. 37, p. 1436-1445, 2008.

[6] SAFARZYNSKA, K.; FRENKEN, K.; VAN DEN BERG, J. Evolutionary theorizing and modeling of sustainability transitions. Research Policy, v. 41, p. 1.011-1.024, 2012.

[7] STEINER, J.; CASSIM, M.; ROBAZZI, A. C. Parques Tecnológicos: Ambientes de Inovação. São Paulo: Instituto de estudos avançados da Universidade de São Paulo, 2008.

[8] Geels F. Understanding system innovations: a critical literature review and a conceptual synthesis: IN: ELSEN, B.; GEELS. F. W.; GREEN, K. System innovation and transition to sustainability: theory, evidence and policy. Part I. USA: E. E. Publishing Ltd, Massachusetts, 2004.

[9] GEELS F. Ontologies, socio-technical transitions (to sustainability), and the multi-level perspective. Research Policy, v. 39. n. 9: p. 495-510, 2010.

[10] STAKE, R. Case Studies. In: DENZIN, N.; LINCOLN, T. Handbook of Qualitative Research. London: Sage, 2005.

[11] Cresweel, John W., 2007, Qualitative Inquiry & Research Design .Choosing Among Five Approaches, Sage Publications, Thousand Oaks/London.

[12] Geels F.Reconceptualising the co-evolution of firms-in-industries and their environments: Developing an inter-disciplinary Triple Embeddedness Framework. Research Policy, v. 43, n. 2: 261-277, 2014.

[13] FPTI. Relatório de Resultados. Itaipu: 2014.

[14] FPTI. Planejamento Estratégico Fundação Parque Tecnológico Itaipu, 2014b. Disponível em:<http://www.pti.org.br/system/files/planejamento_estrategico_fpti_2014_-_2024_aprovacao_final.pdf.> Acesso em: 02 fev. 2016.

[15] MENDONÇA, A. T. B. B. O processo de Transição Sociotécnica para a Eco-Inovação a partir da Relação Multinível: O Caso dos Programas da Itaipu Brasil. 2014. Tese (Doutorado em Administração) ,Universidade Federal do Paraná. Curitiba: 2014.

[16] CIBIOGÁS. Quem Somos. Disponível em:<https://cibiogas.org/quem_somos.> Acesso em: 19 de jul. 2016.

[17] ITAIPU. Projeto Hidrogênio. 2016. Disponível em:< http://www.pti.org.br/projeto-hidrogenio.> Acesso em: 28 jul. 2016.

[18] SCHOT, J.; GEELS, F. Strategic niche management and sustainable innovation journeys: theory, findings, research agenda an policy. Technology Analysis & Strategic Management, v. 20, n. 5, p. 537-554, 2008.

[19] Grin, J. Transitions to sustainable development: new directions in the study of long term transformative chage. New York: Routledge, 2010.

[20] GEELS, F.; KEMP, R. Dynamics in socio-technical systems: Typology of change processes and contrasting case studies, Technology in Society, v. 29, n.4, p. 441-455, 2007.

[21] DOLATA, U. The transformative Capacity of New Technologies: a theory of sociothecnical change. New York: Ed, Routledge, 2013.

[22] KEMP, R.; ROTMANS, J. The management of the co-evolution of technical, environmental and social systems. In: WEBER, M.; HEMMELSKAMP, J. (Eds) Towards Environmental Innovation Systems. Berlin: Springer, 2010.

Chapter 4

Dynamic Extended Exergy Analysis of Photon Enhanced Thermionic Emitter Based Electricity Generation

Canberk Unal, Emin Acikkalp and David Borge-Diez

Abstract

Exergy is the very useful tool to evaluate energy systems besides energy analysis based on the first law of the thermodynamics. In contrast to energy, exergy is not conserved and always decreases. There are many types of exergy analysis involving exergoeconomic, exergoenvironmental, advanced exergy-based analyses, extended exergy analysis etc. In this study, an application of the extended exergy analysis is performed. In extended exergy analysis, not only energy related system is considered but also all materials and energy flows' exergy, non-energetic and immaterial fluxes (capital, labor and environmental impact) are turned into exergy equivalent values and utilized in the analysis, which are calculating for local econometric and social data. These methods can be applied to societies or energy based or non-energy-based system. In this study, dynamic exergy analysis and extended exergy application of electricity generation from photon enhanced thermionic emitter is conducted. According to results, some important values can be listed as; extended exergy destruction, conventional based exergy destruction, extended exergy efficiency, conventional exergy efficiency, extended sustainability ratio, conventional sustainability ratio, extended exergy-based depletion ratio and conventional exergy-based depletion ratio are 542106006 MJ, 542084601 MJ, 0.01094, 0.01094, 1.011, 1.011, 0.978 and 0.989 respectively.

Keywords: exergy analysis, extended exergy analysis, photon enhanced thermionic emitter, dynamic performance evaluation, solar energy

1. Introduction

Energy is a concept transferred from an object to another in form of work or to heat. Energy is a conserved quantity; the law of conservation of energy states that energy can be converted in form, but not created or destroyed [1]. From an economic and social perspective, energy is the most important factor that ensures progress in world living standards and country development. Together with the great developments and changes in the industrial field, the increase in the world population at the same time reveals the need for energy [2]. 87 percent of the energy produced in the world is provided by fossil fuels, 6 percent from renewable sources, and 7 percent from nuclear energy sources. "About 64.5% of the world's electrical energy production is realized by fossil resources (38.7% coal, 18.3% natural gas, 7.5% oil)" [3]. Turkey has no significant energy resources in petroleum and natural gas reserves. In addition, it is a market country that has an important place between Europe and Asia. About 27% of current energy needs in Turkey are known to be met by domestic energy production. When the distribution of Turkey installed power in the energy sector on the basis of resources is examined, natural gas 26.5%, hydraulic energy 32%, coal 21.3%, wind energy 7.7%, solar energy 5.3%, geothermal energy 1.4%, 5.8% share belongs to other resources [4].

The total monetary size of Turkey's energy market is around 84 billion dollars. 60.1 billion dollars of this amount was imported. Renewable energy is the biggest resource that can close a deficit of approximately 24 million dollars. Solar Energy is the biggest renewable energy source. Turkey is a country rich in solar energy [5]. In terms of number of installed solar power plant in Turkey is 556 pieces. Turkey's solar energy installed capacity of 5095 MW in 2018. Turkey ranks 12th in the world in terms of installed capacity [6]. Solar energy is nowadays used in air conditioning (heating–cooling) of residences and workplaces, cooking, supplying hot water and heating swimming pools; in agricultural technology, greenhouse heating and drying of agricultural products; In industry, solar cookers, solar furnaces, cookers, salt and fresh water production from sea water, solar pumps, solar cells, solar pools, heat pipe applications; It is used in a controlled manner in transportation-communication vehicles, signaling and automation, electricity production [7].

Photon enhanced thermionic emission (PETE) combines photovoltaic and thermionic effects into a single physical process to take advantage of both the high per-quanta energy of photons, and the available thermal energy due to thermalization and absorption losses [8]. It can be described by a simple three-step process: first, photoexitation of the valence electrons into a conduction band. Second, thermalization and diffusion of the conduction electrons throughout the cathode; and finally emission of the thermalized electrons into a vacuum and collection by the anode [9]. PETE cells have an advantage in efficiency over purely thermionic cells because the presence of electrons in the conduction band from photovoltaic effects reduces the effective work function of the material, making it easier for electrons to escape into the vacuum. Similarly, PETE cells do not suffer from the problem of low efficiency at high temperatures as standard photovoltaic cells do. This is because the degradation effect is specific to the two-layer semiconductor junction design of standard photocells. Further, since a PETE cell can operate efficiently at high temperature, it can be run in conjunction with a heat engine attached to the anode [10].

Exergy is the property of the system, which is the maximum work potential cand be distracted from a system, once it reaches to equilibrium state with a reference state [11]. Unlike energy, exergy is not conserved, and it's perpetually depleted due to irreversibilities (entropy generation). Some of the exergy is destroyed due to irreversibilities at within the system, and a few of it's thrown into the surroundings from the system boundaries (loss of exergy) [12]. If the exergy losses or destruction decrease, in other words, if the exergy efficiency increases, resource consumption and loss exergy emissions within the method can decrease inversely [13].

Extended Exergetic Accounting (EEA), provides a route to formally convert immaterial and non-energetic commodities into exergetic equivalents [14, 15]. According to EEA, material, energy carriers and externalities (capital, labour, and environmental remediation) represent resource expenses and are expressed in exergy as a unified metric. EEA has incorporated some elements of preexisting theories such as: cumulative exergy analysis, thermoeconomics and life cycle analysis, etc. and combines them into a consistent and expanded formulation (Extended Exergy) [16]. There are fundamental similarities between EEA and exergy methods. Like thermoeconomics, EEA results in a system of cost equations in which though inhomogeneous quantities like labour, material and energy flux, capital are all homogeneously expressed in primary exergy equivalents. Like Cumulative Exergy Consumption, EEA computes the cumulative primary exergy "embodied" in a product over its entire production process. Like Exergy Life Cycle Assessment, EEA computations cover the entire life cycle of the considered system [17].

2. System description

Electricity generation by means of Solar source is dived into two part as: directly, which photovoltaic panels is used, and using concentrated solar collectors to obtain heat energy as heat input in a heat engine. A photon enhanced thermionic includes these two principles. Its emitter is made of a semiconductor as cathode and there is a vacuum gap between anodes. Considered system is illustrated in the **Figure 1** where a solar concentrator having 1000 concentration rate. Operation principal of the PETE can be expressed in next sentences. Solar energy passes into concentrator and illuminate the cathode heated up. This energy causes to excite electrons and electron population increase at conduction band. They have bigger energy than electron affinity and emitted into vacuum. Released electrons are collected by the anode and this is resulted in currency. The higher cathode temperature causes the more electron emittance and increase the efficiency of the process and this is the reason that its electricity efficiency is higher than the conventional photovoltaics (PV) panels. Another advantage of it is that it can be heated up via waste heat from an engine, turbine, and industrial heat to generate electricity when sun is no exist in the nighttime. In this paper GaAs is selected as cathode material since it is very promising candidate.

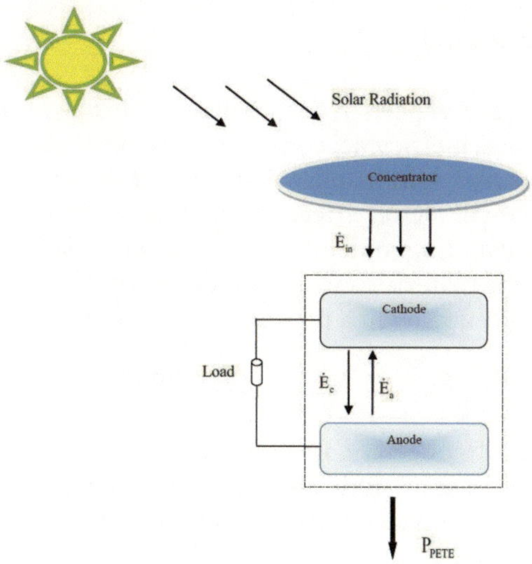

Figure 1.
Schematic of the system.

3. Methodology

Mathematical description and explanations of physical meaning of the exergy, extended exergy accounting and photon enhanced thermionic emitter are described. Daily dynamic and annual analyses are performed by TRNSYS and

Simulink and analyses are conducted. Izmir, which is the third biggest city in Turkey and has great monetary flow, is chosen.

3.1 Exergy analysis

Exergy is one of the most useful method used for evaluating the performance of various thermodynamic systems because exergy destruction is a measure of the losses in the system. Losses in the system is called as irreversibilities that is resulted from the entropy generation. These losses are main reason of inefficiencies and excessive depletion of the fuel source. In other words, exergy is the measure of how quality energy resource is used, and it is not conserved like energy and always decrease. Because of these, exergy analysis should be utilized in optimization and design studies of energy conversion devices. Exergy analysis provided to determine place and amount of the irreversibilities, and it must be decreased to save energy, money and emission by obtaining more efficient systems. Conventional exergy analysis involves three variables, $\dot{E}_F$, $\dot{E}_P$, $\dot{E}_D$, which represent the fuel exergy rate, the product exergy, and the exergy destruction rate, respectively. Exergy is not conserved, in contrast to energy.

The exergy destruction rate can be calculated as follows.

$$\dot{E}_D = \dot{E}_F - \dot{E}_P \tag{1}$$

The exergetic efficiency is

$$\varphi = \frac{\dot{E}_P}{\dot{E}_F} \text{ or } \varphi = 1 - \frac{\dot{E}_D}{\dot{E}_F} \tag{2}$$

3.2 Extended exergy analysis

Exergy analysis is very important tool to describe losses resulted from the irreversibilities, which cause inefficiencies, as it is mentioned above. However, exergy analysis is mostly used as a sustainability indicator, it does not include material or resource depletion, human factor, effect of capital and environmental remediation. In extended exergy accounting, not only not only energy flows are included but also exergy equivalent of the materials, resources, the environmental remediation costs and labor and capital needed by in a process. *EE* can be expressed as follows:

$$E_{in} + E_L + E_C + E_E = EE\ (MJ) \tag{3}$$

E_C indicates the total monetary cost of the equipment, expressed in terms of its equivalent exergetic content, E_L indicates the sum of the labor contribution expressed in terms of its equivalent exergetic input, and E_E is exergy consumption of environmental remediation for removing pollutant emissions. The difference of extended exergy accounting to other exergy analysis approaches is the inclusion of labor, capital, and environmental impact as exergy units. 'The value of 'ee_L' is calculated as.

$$ee_L = \frac{365 N_h e_{surv} HDI}{HDI_o N_{wh}}\ (MJ/wh) \tag{4}$$

ee_L' means the specific exergetic equivalent of the labor. 'E_{surv}' refers to the minimum exergy consumption for a person's survival. 'N_h' is the value of population. 'N_{wh}' refers to the number of working hours per year. f is the consumption correction factor for modern living standards.

$$f = \frac{HDI}{HDI_0} \tag{5}$$

'HDI' is the human development index published by the United Nations every year. 'HDI_0' is the human development index of a pre-industrial society. E_L' is the total exergy equivalent of Labor and represents contribution of the human work on the process. 'E_L' is found by multiplying the specified exergetic equivalent of labor by the number of working hours per year.

$$E_L = ee_L N_{wh}\ (MJ) \tag{6}$$

'ee_C' is the specific exergetic equivalent of the capital. The value of 'ee_C' is calculated as.

$$ee_C = \frac{365 e_{surv} N_h HDI}{HDI_o S}\ (MJ) \tag{7}$$

'S' indicates the amount of the national monetary wage as the average wage and the total energy equivalent of the currency.

The value of 'E_C' is the amount of monetary flow in the process and it is calculated as:

$$\mathrm{E}_C = \mathrm{ee}_C \sum CC\ (MJ) \tag{8}$$

where, 'C' is the capital cost. The first thing in calculations is to collect data involving population of the country, numbers of the workers in the country, average annual wage, human development index, capital circulation of the country. Second thing is to calculate number of consumed materials, releasing CO_2, total exergy input, exergy for the survival, exergy equivalents of the specific labor and capital are done. Using previous values, total exergy of the equivalent labor and capital, cumulative exergy consumption, product exergy (or equivalent) and exergy destruction, environmental remediation can be obtained. In this paper, only the operations calculations of the considered system are researched. Therefore, cumulative exergy and environmental remediation are excluded, which is no emission realizing in operation conditions.

3.3 Analysis of the PETE

Photon powered thermionic emission (PETE) is a new concept of solar power generation. It combines different quantum and thermal mechanisms directly in a physical process. It is possible to overcome both the disadvantages faced by conventional thermal systems and the natural loss of photovoltaic cells.

Unlike traditional photovoltaic cells, the PETE has higher efficiency at high temperatures, which gives an opportunity usage as bottom cycle in hybrid applications for efficiency increasing.

It sees the cathode as a clumped system with average properties and no spatial variation, so it can be called a zero-dimensional model. The cathode absorbs all band gap radiation. The anode is metallic and has good reflectivity and the void charge in the gap between the electrodes is ignored.

The output power current is given by;

$$P_p = (j_c - j_a)V \tag{9}$$

'J_c' is the density of the emission current from the cathode surface and 'J_a' is the density of the emission current from the anode surface. 'V' is the cathode electron emitting area.

The emission current density of the cathode is proportional to the Electron concentration 'n'. The emission current density of the cathode can be expressed as:

$$j_c = en\sqrt{\frac{kT_c}{2\pi m_e}}e^{-\chi/kT_c}\ (A/cm^2) \tag{10}$$

Where 'n' is the conduction band electrons concentration. 'm_e' is the electron effective mass. 'χ' is the electron affinity. 'T_C' is the anode temperature. 'k' is Boltzmann's constant.

The reverse emission current density from the anode follows the standard thermionic emission formulation:

$$j_a = A_o T_a^2 e^{-\Phi_a/kT_a}\ (A/cm^2) \tag{11}$$

'A_0' 120 A cm^{-2} K^{-2} is the Richardson–Dushman constant. 'T_A' is the anode temperature. 'k' is Boltzmann's constant. The energy barrier for emission from the anode is equal to the anode operating function '$Ø_a$' for voltage above the flat band value.

$$\Phi_c = \chi + E_g - E_f \tag{12}$$

4. Results and discussion

In this paper, performance and sustainability of the photon enhanced thermionic emitter is evaluated via extended exergy analysis. This analysis is performed for annual and hourly values. In **Tables 1–3**, annual results can be seen, and hourly values are shown in **Figures 2–8**.

Annual values are arranged in tables. In this tables, some calculated values are put in order as e_{surv} is the daily exergy amount needed a person for surviving, ee_L is the specific exergy equivalent of the labor, which represents daily labor exergy per

E_{in} *(MJ)*	N_h -	N_w *(h)*	N_{wh} *(wh)*	*S (Euro/ year)*	M_2 *(MEuro)*	*f*	e_{surv} *(MJ/day-person)*	ee_L *(MJ/ wh)*	ee_K *(MJ/ Euro)*
588582	83900373	33810000	79115400000	11325	460916.40	14.40	10.50	42.68	12.09

Table 1.
Data used in extended exergy accounting.

EE_L *(MJ)*	EE_K *(MJ)*	E_P *(MJ)*	E_D *(MJ)*
2133.75	1551.24	353513.77	547732043

Table 2.
Results of the extended exergy accounting.

φ	SI	y
0.597	2.481	0.252

Table 3.
Extended exergy evaluation indices.

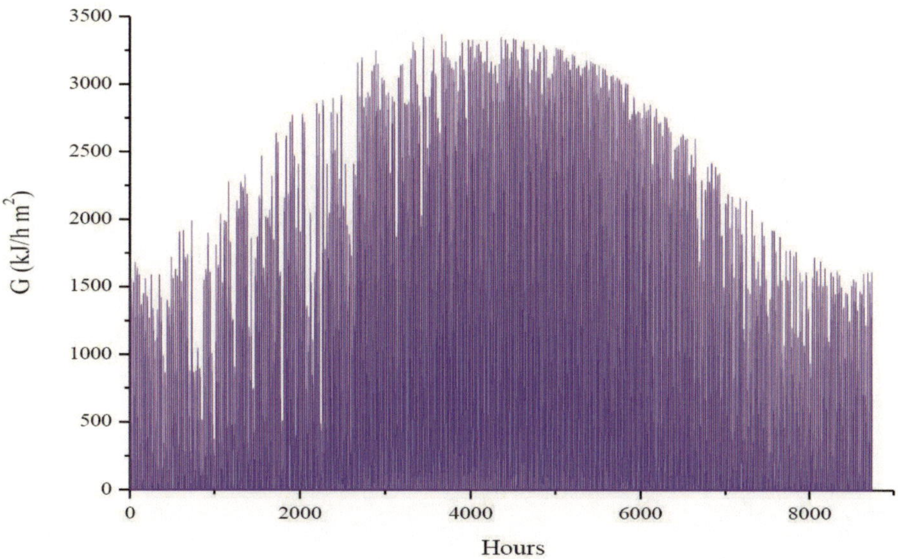

Figure 2.
Solar irradiation.

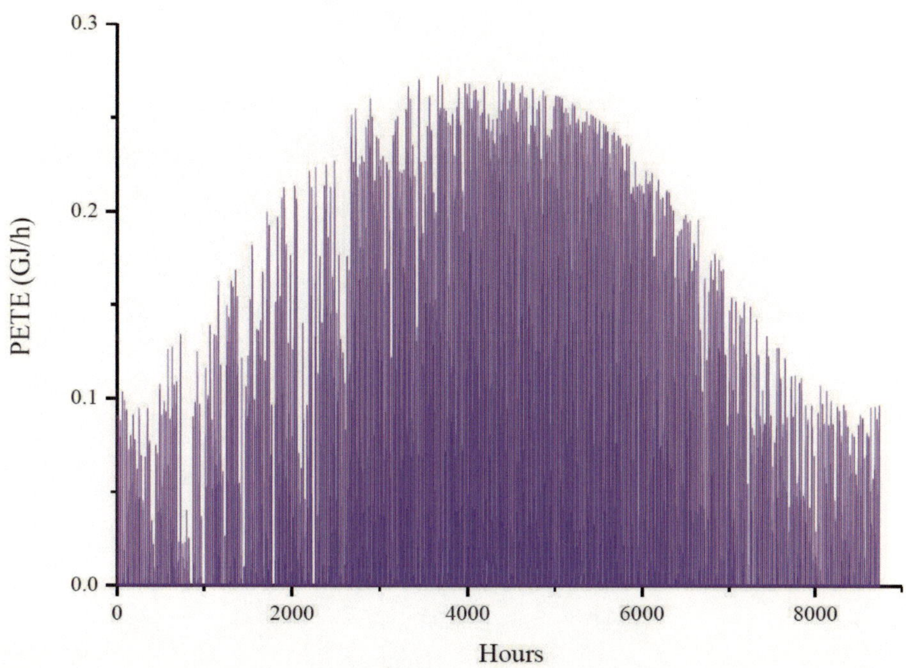

Figure 3.
Exergy output rate.

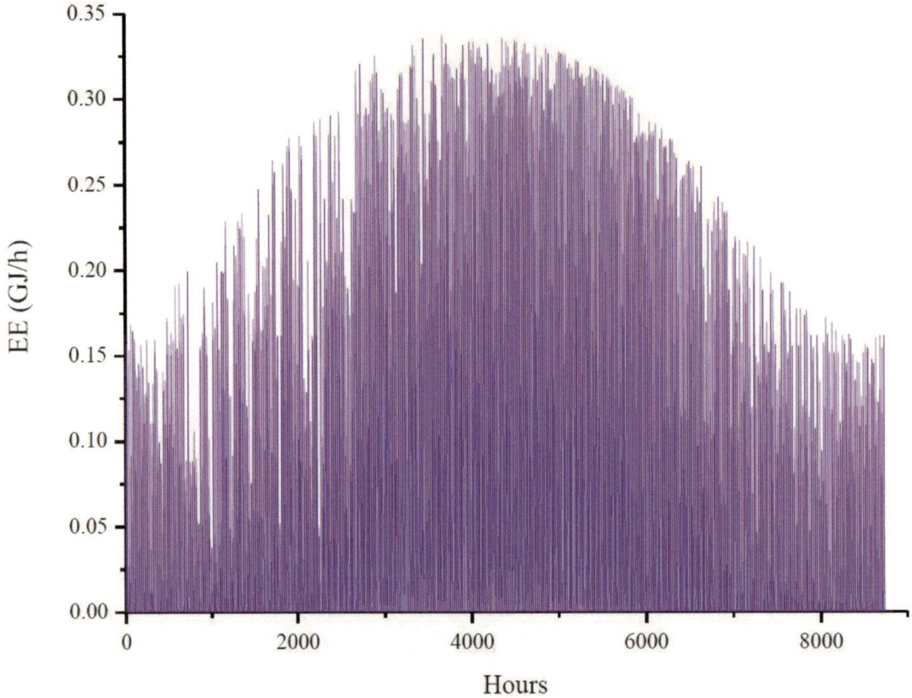

Figure 4.
Extended exergy rate.

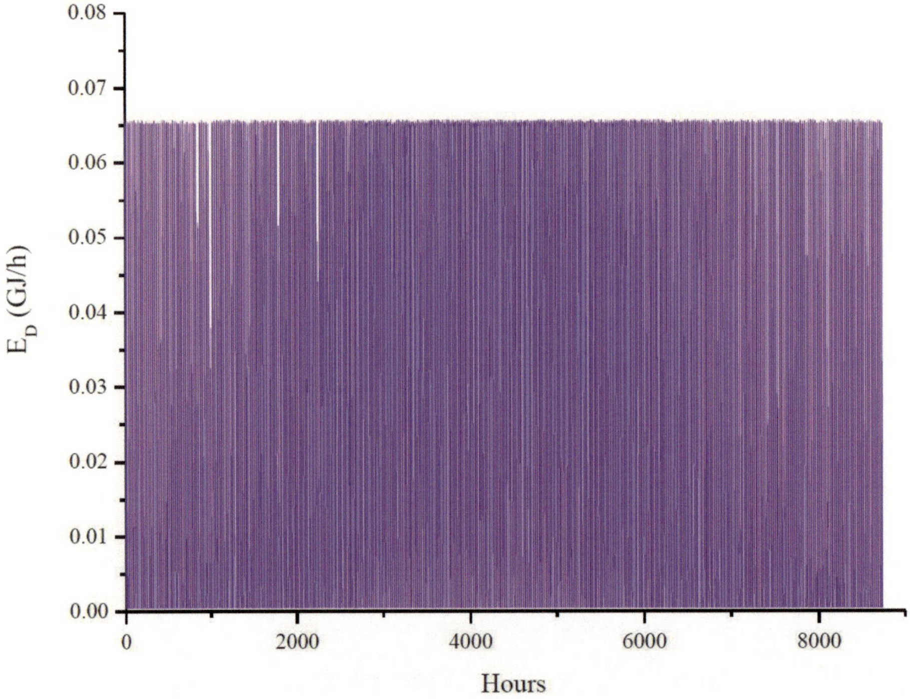

Figure 5.
Exergy destruction rate.

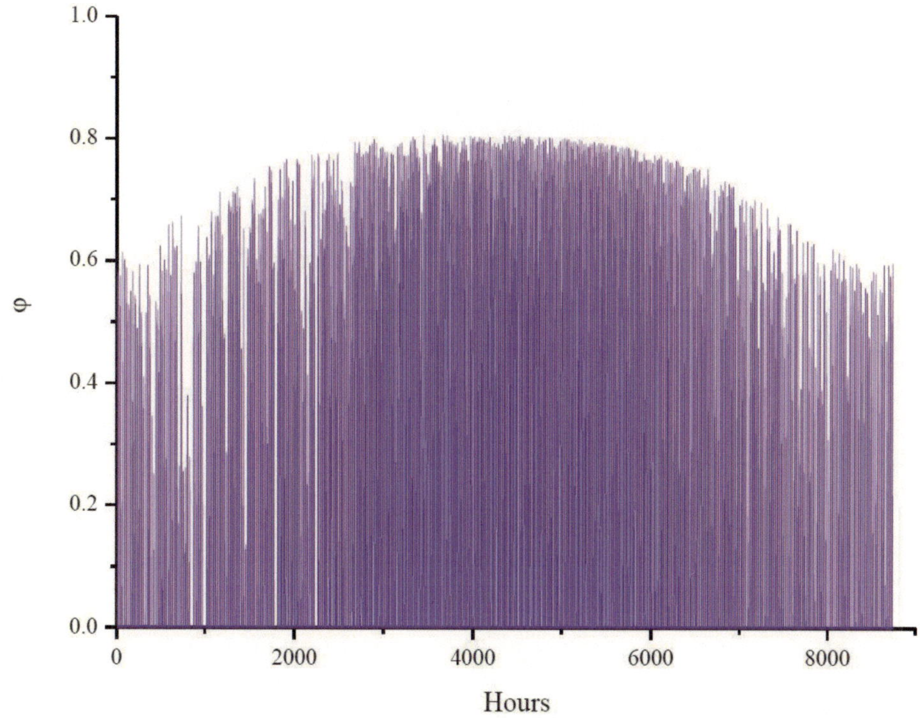

Figure 6.
Extended exergy efficiency.

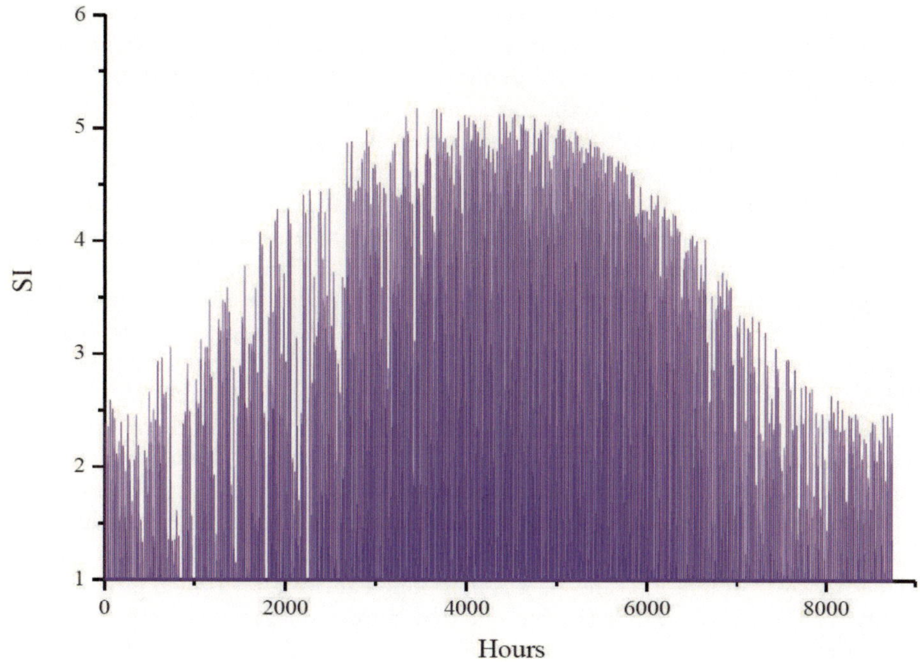

Figure 7.
Sustainability index.

workhour, ee_K is the specific exergy equivalent of the capital donating exergy per Euro, EE_L, EE_K, E_P, E_D, φ, SI, y, are the total equivalent exergy of the labor, total equivalent exergy of the capital, product exergy aiming output from any system,

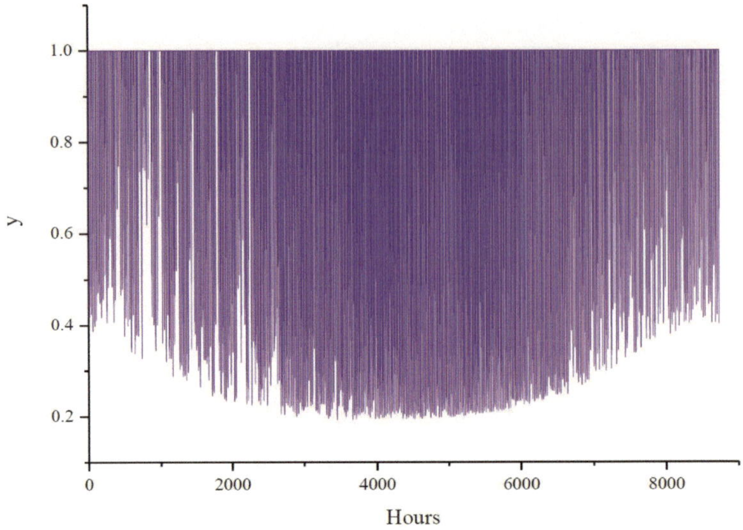

Figure 8.
Exergy destruction ratio.

exergy destruction is depleted exergy in any system, exergy efficiency, sustainability index and exergy destruction rate respectively. According to results, corresponding values are equal to 10.50 *MJ/day person*, 42.68 *MJ/wh,* 12.09 *MJ/Euro,* 2133.75 *MJ,* 15511.24 *MJ,* 353513.77 *MJ,* 238753 *MJ,* 0.60, 2.48 and 0.40. Examining performance indices, extended exergy efficiency which is measure of how quality energy used for product is 0.60, *SI* is the indicator how sustainable a system and equal to 2.48 and *y* is the depletion rate of the source. According to these results, system is 60% closer to the ideal one, *SI* is the 1 for the non-sustainable system and this means considered system 2.48 times sustainable. Last index shows that 0.25 of the exergy resources is depleted. Extended exergy is accounted as 592267.04 *MJ,* share of the exergy equivalent of the labor and capital can be neglected, since they are much lower than 1%. These means capital and labor has no effect on the considered system.

Figure 2 represents the solar irradiance, which is the solar energy input, of the Izmir where maximum irradiation rate is 3370 $kJ/h\ m^2$. Daily values of the PETE are depicted in **Figure 3**, it reaches to 272234.01 *MJ,* while its average value is 80923.38 *MJ*. Sum of the daily equivalent exergy of the labor and capital (418.82 *MJ*) is nearly 0.6% of the product exergy, which indicates these have no important effect on the product. It is indicator that capital and the labor are not important, however, technology is the most important factor.

Figure 3 is about the exergy output which means change of the electricity generation at the system. Naturally, electricity output in summer times because of the higher solar irradiation rate. As it is mentioned above total annual product exergy is 353513.77 *MJ.* When average and maximum values are researched, one can see that average is 80914.11 *MJ* and the maximum value is 272234.01 *MJ.* These values represent aimed product, and it is claimed as big as possible.

Figure 4 shows daily changes of the extended exergy and maximum value is equal to 337481.82 *MJ* and average is 135696.79 *MJ*. This is made of the sum of the equivalent exergy of the labor, capital and exergy input, which is the solar energy input to the system. However, it is mostly made of the input exergy. This means that solar exergy is the biggest part of the extended exergy accounting.

Similarly, **Figure 5** is about the change of the exergy destruction in which the maximum is 65724.17 MJ and average is 54773.42 MJ. In results, it is seen that exergy destruction rate is equal the sum daily equivalent exergy of the labor and capital when there is no production like extended exergy.

In **Figures 6–8**, daily results of the exergy efficiency, SI and y are indicated. Their maximums are 0.81, 5.17 and 1 respectively. Their expression can be done as follows. Quality of using energy source is 81%, sustainability of the system can reach to 5.17 and depletion of the energy source rises to 1, which means all the exergy is depleted. Their average values are exergy efficiency, SI and y is the 0.20, 1.62 0.80, like explained above, quality of the energy usage is only 20%, while sustainability ratio is relatively low and exergy source is 80% depleted. These results show the most important reason of this that PETE is relatively new technologies and they have not enough efficiency despite their promising and developing technologies.

According to the results, share of the exergy equivalent of the labor in the extended exergy accounting can be neglected because of its very small values. This means that effect of the labor has no important effect. Similarly, the rate of the exergy equivalent of the capital the, i.e., exergetic reflection of the monetary flow in the extended exergy accounting can be neglected too. This reason might be expressed as PETE technology is relatively cheaper than other renewable energy technologies. Another, important point is that the exergy destruction is very great and equal to nearly 30–40% of the extended exergy. This means that 30–40% of the input exergy cannot be converted to product and it is dissipated. Another important approach is to compare product exergy to exergy equivalent of the labor and capital. Therefore, one can determine how labor and capital are consumed for producing one-unit product. If these are researched, it is seen that consumed labor exergy is nearly 0.6% of the product exergy. This can be interpreted as labor has no important effect to produce electricity. It is expected results since PETE does not require any intervention to operate and only labor force is need to maintenance process. Similar investigation should make for the capital and results shows that influence of the capital on the product is the lower than the labor's one. Rate of the capital to the per unit product is calculated as 0.4%. As it is mentioned above, PETE is relatively cheaper than other renewable energy technologies and maintenance costs are not so expensive science that are relatively simple and do not have operating parts.

5. Conclusion

In this paper, solar energy usage by means of photon enhanced thermionic emitter is analyzed via extended exergy analysis, which is one of the methods for evaluating sustainability, and dynamic performance analysis is conducted. Izmir, which is the third biggest city in Turkey and has great monetary flow, is chosen as considered location. Since, it has good solar potential. Some important results and recommendations are listed as follows:

- Although system has relatively low exergy efficiency besides the high exergy destruction rate, it can be tolerable. Since, main energy source is solar power, which is renewable, and solar energy has no environmental harmful.

- Exergy equivalent of the labor has bigger than the exergetic equivalent of the capital. This difference is resulted from the higher value of the specific exergy equivalent of the labor, in other words, exergy provided by labor are valuable than capital.

- The maximum values of the indices show that it has good potential where global irradiation is high and generally, its losses are greater for the times when the solar irradiation is lower.

Finally, it is recommended that extended exergy analysis should be applied to other systems, which are energy conversion systems, because of its advantages in sustainability aspects mentioned before.

Author details

Canberk Unal[1], Emin Acikkalp[2*] and David Borge-Diez[3]

1 Mechanical Engineering Department, Engineering Faculty, Bilecik S.E. University, Bilecik, Turkey

2 Department of Mechanical Engineering, Engineering Faculty, Eskişehir Technical University, Eskişehir, Turkey

3 Department of Electrical, Systems and Automation Engineering, University of León, Spain

*Address all correspondence to: eacikkalp@eskisehir.edu.tr

References

[1] Energy, Acces Date; (Oct 2020) https://en.wikipedia.org/wiki/Energy

[2] Koç, A., Yağlı, H., Koç, Y., Uğurlu, İ., Genereal Evaluation of Energy Outlook in Turkey and the World, (In Turkish), Engineer and Machinery. 2018; 59, 86-114.

[3] Zaimoğlu, Z., Effects of Energy Consumption Growth in Turkey, (In Turkish), İstanbul University, Social Sciences Institute, Master Thesis, 2019.

[4] Çıray, B., Energy Policy in Turkey and Incentives Provided to Renewable Energy Production, (In Turkish), Balıkesir University, Economics Department, Master Thesis, 2019.

[5] Altuntop, N., Erdemir, D., Development of the Solar Energy in Turkey and Around the World, Mühendis ve Makina, 2013; 54, 69-77.

[6] Erdoğan, N., Interaction Between the Reflections and Financial Incentives for Renewable Energy and Renewable Energy Production in Turkey, (In Turkish), Sivas Cumhuriyet University, Social Sciences Institute, Master Thesis, 2020.

[7] Varınca, K.B., Gönüllü, T. M., Solar Energy Potential and Potential Use of this Degree in Turkey, Method and A Research on Prevalence, UGHEK, 2006; 270-275

[8] J.W. Schwede, I. Bargatin, D.C. Riley, B.E. Hardin, S.J. Rosenthal, Y. Sun, et al., Photonenhanced thermionic emission for solar concentrator systems, Nat. Mater. 2010; 9, 762–767

[9] Xiao, G., Zheng, G., Ni, D., Li, Q., Qui, M., Ni, M., Thermodynamic Assessment of Solar Photon-Enhanced Thermionic Conversion, Applied Energy. 2018; 223, 134-145

[10] Photon-Enhanced Thermionic Emission, Acces Date; Oct 2020, http://large.stanford.edu/courses/2010/ph240/brown-cohen2/

[11] Koroneos C., Spachos T., Moussiopoulos N., 'Exergy Analysis Of Renewable Energy Sources' Renewable Energy, 28 (2003) 295–310

[12] Rahim M. A., Gündüz D., 'Energy and Exergy Analysis of A Gas Turbine Cogeneration Power Plant: Application For Ankara Conditions ', (In Turkish), (2013), 19-27

[13] Tekel E., 'Energy And Exergy Analysis Of Thermal Power Plants ', Pamukkale University, Institute of Science, Master Thesis, 2006

[14] Sciubba, E., Extended exergy accounting: towards an exergetic theory of value. In: Proceedings of the ECOS '99, Japan. 1999, 85-94.

[15] Sciubba, E., Exergo-economics: thermodynamic foundation for a more rational resource use, International Journal of Energy Research. 2005; 29,613–636

[16] Seckin, C,. Sciubba, E., Bayulken, R. A., Extended Exergy Analysis of Turkish Transportation Sector, Journal of Cleaner Production. 2013; 47, 422-436

[17] Rocco, M.V.,Colombo, E., Sciubba, E., Advances in exergy analysis: a novel assessment of the Extended Exergy Accounting method, Applied Energy. 2014; 113, 1405-14

Section 2

Physics and Biology of Renewable Energy

Chapter 5

Scale Invariant Turbulence and Gibbs Free Energy in the Atmosphere

Adrian F. Tuck

Abstract

A method of calculating the Gibbs Free Energy (Exergy) for the Earth's atmosphere using statistical multifractality — scale invariance - is described, and examples given of its application to the stratosphere, including a methodology for extension to aerosol particles. The role of organic molecules in determining the radiative transfer characteristics of aerosols is pointed out. These methods are discussed in the context of the atmosphere as an open system far from chemical and physical equilibrium, and used to urge caution in deploying "solar radiation management".

Keywords: scale invariance, entropy, exergy, atmosphere

1. Introduction

Earth's atmosphere is an open system far from equilibrium, both physically – there is vigorous circulation, and chemically – for example, the methane/oxygen ratio is some 30 orders of magnitude larger than the equilibrium value. It is not isolated; there are varying fluxes of photons in and out, water exchanges on a 10-day time scale with the oceans, carbon dioxide, many organic molecules, nitrous oxide, methane, methyl chloride, ammonia and sulphurous compounds are subject to biogeochemistry at both land and sea surfaces. Long-lived molecules such as chlorofluoromethanes (50–100 years) can be re-emitted from both land and sea, and even the major constituents, O_2 (order 10^4 years) and N_2 (order 10^6 years) are cycled by geochemistry, biochemistry and lightning. Last but not least, condensed matter in the form of aqueous aerosols is produced by gas to particle conversion, by clouds and from sea spray, and which serve as condensation nuclei for cloud droplets, ice crystals and precipitation. The air-water interface is an important reaction venue [1], and often accelerates reactions [2–6]. Some or all of these molecules and particles affect the transmission of both solar visible and ultraviolet and terrestrial infrared radiation, and consequently are central to the maintenance of atmospheric temperature.

Since the atmosphere is so far from equilibrium, standard statistical thermodynamics calculable by quantal chemical techniques are not applicable, either in the air or aerosols. The thermodynamic formulation of statistical multifractals has been shown to be a mapping, not just a formal coincidence by Lovejoy and Schertzer [7], following their analyses of atmospheric variables as statistical multifractals [8, 9]. It has been used to demonstrate, by analysis of observations, that the current global heating is attributable almost wholly to carbon dioxide emissions from fossil fuel burning [10]. Note that this is a numerical model-independent result. By combining these analyses with results from molecular dynamics calculations [11] it has been shown that atmospheric turbulence is an emergent property of molecular gas populations in an asymmetric environment, scale invariant over the 15 orders of magnitude in length scale, from the mean free path at surface pressure to a great circle [5, 12, 13]. The scale invariance approach to atmospheric thermodynamics has been successfully applied to aircraft data from the stratosphere [14] and combined with theoretical and experimental results for aerosol particles to form a potential pathway to understanding and exploiting the observed acceleration of many chemical reactions, some of them atmospheric chemical, in and on microdroplets compared to their slower or non-occurrent behaviour in bulk fluid [5]. Microdroplet surfaces are where free energy is concentrated, even in a pure homogeneous droplet, as surface tension. So, for both gaseous composition chemistry and the aerosol population, analysis can be attempted with this approach. All these factors are directly relevant to the calculation of global heating under anthropogenic perturbations. The issue is examined in the rest of the chapter by this scale invariant thermodynamics approach. A useful review of energy production and its interaction with the environment is given in Goede et al. [15]. Further considerations are in Refs. [16–18].

The above approach provides perspective on geoengineering by "solar radiation management" when combined with the complexities of aerosol particles in the lower stratosphere [19]. Because of dissipation by infrared radiation to space, any method of reducing global heating by fossil fuel burning cannot be powered by combustion of coal, oil and gas: renewable sources must be deployed.

2. Atmospheric entropy and Gibbs free energy from scale invariance

This account draws on those in Tuck [5, 12–14, 17, 19] and references therein. In order to maintain scale invariance, the same processes must be at work scale by scale over the range concerned, 40 metres to a great circle (40,000 km) in this case. That process is manifest as turbulence, driven by the thermodynamic need to minimise Gibbs free energy, which it does by dilution of energy density. **Figure 1** shows an example of an airborne temperature trace at 200 m resolution. Analyses for the scaling exponent H and the intermittency C_1 are also shown, see below for definitions.

Satellite data have extended the aircraft data's range from an Earth radius to a circumference [16]. A view of the atmospheric circulation then emerges wherein organised flow, the general circulation, is driven by turbulence that enables the minimisation of Gibbs free energy. That free energy arises from the absorption of photons from the low entropy solar beam by ozone, some clouds, a little by water vapour and by the surface. The necessary dissipation is achieved by infrared radiation to space from water vapour, carbon dioxide, ozone, water dimers, nitrous oxide, methane and assorted halocarbons. Radiation of both solar and terrestrial photons is by gases whose fluctuating abundance is modulated by atmospheric turbulence. The radiative effect of aerosols and clouds, including absorption, emission and scattering is an important component in the radiative balance, and the abundances of both are also modulated by turbulence.

Why turbulence? Is a question addressed successively by Horace Lamb, Théodore von Kármán and Werner Heisenberg, all of whom anticipated a disappointment in or hope for divine enlightenment upon entry to heaven – possibly with a Shakespearean degree of irony. Richard Feynman described it as the last unsolved major problem in classical physics. Here the view is put forward that it is an emergent property of

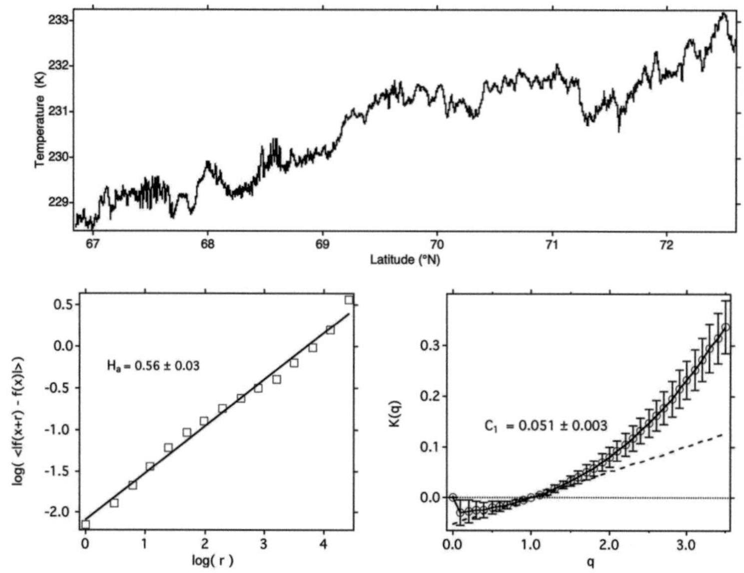

Figure 1.
Temperature on an approximately horizontal flight leg of 4700 seconds from (65°N,148°W) on a great circle to the NE, on 19970506 (yyyymmdd). The data are averaged to 5 Hz, or about 40 m horizontal resolution (top). Lower left, log–log plot to determine Hurst exponent H_1*; the value of 0.56 ± 0.03 confirms the theoretical value of 5/9 predicted by statistical multifractal theory (the flights combine both horizontal and vertical motion). Lower right, intermittency* $C_1(T)$*, which correlates with the mean T for the segment when the data for all suitable flight legs are so plotted.* $C_1(T)$ *is also correlated with the ozone photodissociation rate, see text and* ***Figures 3*** *and* ***4****.*

molecular populations in an asymmetric environment, following the molecular dynamics calculations of Alder & Wainwright [11], as taken up by Tuck [5, 12–14, 17]. In the atmosphere, it is driven by the Gibbs free energy arising from the entropy difference between the incoming organised beam of solar photons and the outgoing less organised flux of infrared photons over the whole 4π solid angle.

The statistical multifractal analysis is briefly outlined as follows.

It has been argued recently that G is computable from observations in a scale invariant medium, and shown to work [5, 14]:

$$G \equiv -\frac{K(q)}{q} \tag{1}$$

where $K(q)/q$ is a scaling quantity related to partition function f, Boltzmann constant k and temperature T by:

$$T \equiv \frac{1}{kq} \tag{2}$$

$$f \equiv \exp\{-K(q)\} \tag{3}$$

The relationship of $K(q)$ to the Hurst exponent H is given in Tuck [14] as.

$$H = H_q + K(q)/q \tag{4}$$

where

$$H_q = \frac{\zeta(q)}{q} \tag{5}$$

Quantity in Eqs. (1)-(5)	Statistical thermodynamics	Scaling equivalent
Temperature	T	$1/qk_{\text{Boltzmann}}$
Partition function	f	$e^{-K(q)}$
Energy	E	γ
Entropy	$-S(E)$	$c(\gamma)$
Gibbs energy	$-G$	$K(q)/q$

Table 1.
Equivalence between statistical thermodynamic and scaling variables.

and $\zeta(q)$ is the linear slope of a log–log plot of the first order structure function of the fluctuations of the observed variable versus a lag parameter covering the range of the variable. The Hurst exponent H, the intermittency C_1 and the Lévy exponent α are the scaling exponents that comprise the statistical multifractal description of atmospheric variability. Their derivation can be found in [14], with their statistical thermodynamic equivalences in **Table 1**.

By applying scale invariance to the bulk medium, and separately to the microdroplet, the ΔG between the two may be obtained by difference. Application of high-resolution experimental and imaging techniques would then enable a comparison with the values obtained by quantum statistical methods applied to the reactant molecules [5]. The technique has been successfully applied to the air itself [14].

3. Some consequences of atmospheric statistical multifractality

As observed, atmospheric variables display the fat-tailed probability distribution functions characteristic of statistical multifractality; the Gaussians associated with Einstein-Smoluchowski diffusion are conspicuous by their absence. An example for temperature is shown in **Figure 2**.

The results from Alder and Wainwright [11] combined with statistical multifractal analysis [5, 12, 13, 17] imply that the air is not at chemical equilibrium and consequently the gas constant R is not sufficient to describe the molecular behaviour of the atmosphere. All current atmospheric models employ the gas constant in this manner. An experiment is suggested to test this point: measure the populations of the rotational energy levels of the major constituents, N_2 and O_2 to see whether they are at equilibrium – populated according to the Boltzmann distribution – or not. Direct measurement of the probability distribution of air molecule velocities would be a significant advance and check on the theory. The temperature should be compared between that consistent with Eq. (2) and that necessary to account for the rotational and translational populations. It is an important point that the implied higher population than Boltzmann equilibrium means that the far wings of the water vapour and carbon dioxide will be stronger than at equilibrium because these wings are caused by the molecular collisions with the highest speeds. That is where there is significant influence in calculations of the atmospheric radiative transfer, because unlike many line centres, they are not self-absorbed. There is observational evidence supporting this view of atmospheric temperature. **Figure 3** shows a plot of the ozone photodissociation rate $J[O_3]$, against the intermittency of temperature, $C_1(T)$, for flights in the lower Arctic stratosphere in the summer of 1997 and in the winter of 2000.

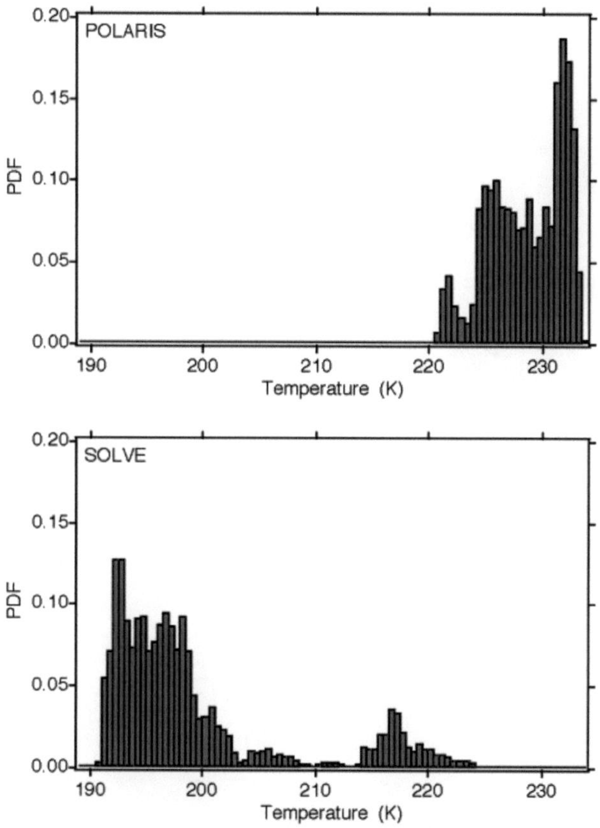

Figure 2.
Probability distribution functions (PDF) of temperature from ER-2 flights at 18–20 km in the Arctic April–September 1997 (top) and January–march 2000 (bottom). The millions of data points are averaged to 1 Hz, corresponding to approximately 200 m horizontal resolution. Note the long or fat tails; Gaussians are not seen.

The correlation is consistent with the effects of translationally hot oxygen atoms from ozone photodissociation being unequilibrated and conveying this to the air molecules as a whole. **Figure 4** shows the average temperature against intermittency $C_1(T)$ for the flight leg, and the correlation persists.

The reason for this, and the theoretical explanation, is in Chapter 5.1 of reference [12] and Section 4.2 of [14]. Further discussion is given in the next section. There are no suitable observational data to apply this analysis in the troposphere.

Now we consider the effects of scale invariance and statistical multifractality on clouds and aerosols. This approach was pioneered by Schertzer and Lovejoy [8, 9] and described at length in the context of clouds and radiation in Lovejoy and Schertzer [7]. Both phenomena pose major problems for current general circulation models (GCMs) that are used to attempt predictions of the future evolutions of climate. Lovejoy [10] has demonstrated that such models do not predict climate, they predict macroweather, the fluctuations on scales from 10 days to 30 years. 10 days is the approximate time for air to circle the globe, and it shows the scaling expected, namely 23/9 dimensional statistical multifractality. On time scales longer than 30 years, the data show similar scaling. On the intermediate 10-day to 30-year scale, the observations scale differently, and constitute the macroweather scale, which is what the so-called climate models predict. Macroweather forecasting of course has potential utility, but it is not climate prediction.

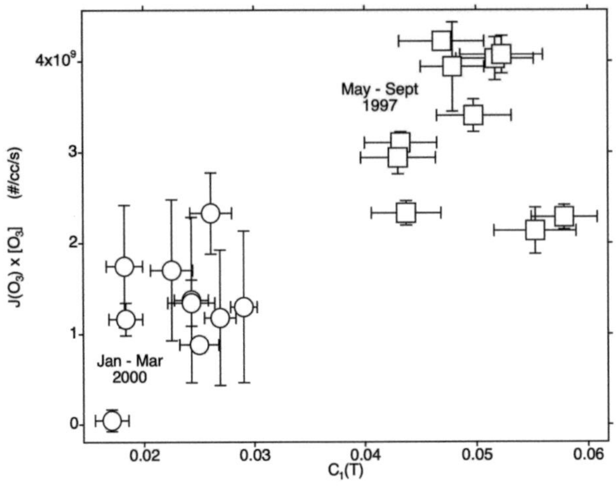

Figure 3.
For all suitable ER-2 flights April–September 1997 and January–march 2000. J[O_3] *on the ordinate,* C_1(T) *on the abscissa. The ozone photodissociation rate is averaged over the flight leg, with vertical bars representing the standard deviation. The intermittency of temperature is derived from the slope of the curve for each flight leg, see* ***Figure 1****, bottom left. The horizontal bars are the standard error of the slope of the line fit.*

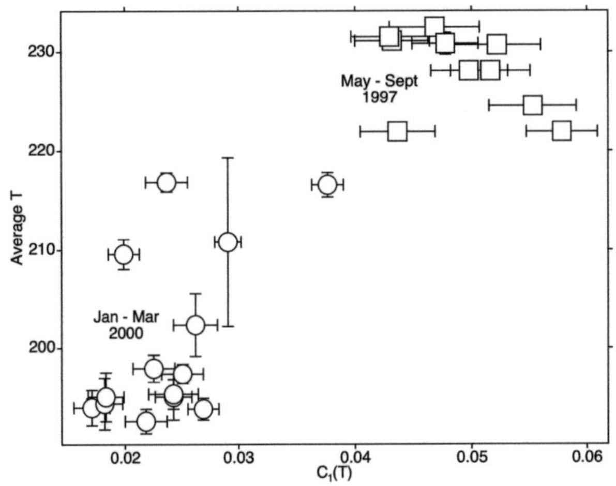

Figure 4.
As for ***Figure 3****, but with* T *averaged over the flight leg on the ordinate. The macroscopic temperature is proportional to the mean square velocity of the air molecules, supporting the suggestion that translationally hot photofragments from ozone photodissociation account for the correlations in* ***Figures 3*** *and* ***4****.*

Clouds are still a major uncertainty in any modelling, because they involve the physical chemistry of all three phases of water, the complex chemistry of gas-to-particle conversion, the role of aerosols of varying sizes in acting as condensation nuclei, particularly as regards the role of organic surfactants. All of these phenomena affect the transmission of radiation, both UV/visible and infrared. The altitude of the clouds is also critical, via their temperature and hence state. As an example, we note that the organic content of lower stratospheric aerosols plays a disproportionately large role in their radiative influence [18]. This whole area of aerosols, clouds and radiation needs examination by scale invariant techniques, from individual particles to cloud decks. The scale invariant Gibbs free energy is, we argue, an appropriate tool. Its effects are of course intimately related to atmospheric temperature and its maintenance via dissipation through intermolecular energy exchange and subsequent infrared radiation to space.

4. Atmospheric temperature

Before considering the role of Gibbs free energy in the atmosphere, we take a detailed look at a sample of evidence from aircraft flights, drawing on a selection of figures from Tuck [12–14, 17, 19] to examine what processes determine atmospheric temperature. The discussion starts with **Figures 1–4**, taken from reference [17], and questions and answers in [20].

Figure 1 shows a temperature trace, and the plots for obtaining the Hurst exponent H_1 and the intermittency C_1 from it. It is a great circle flight in the lower stratosphere during Arctic summer. The value $H_1 = 0.56 \pm 0.03$ is observational confirmation of the theoretical value of 5/9 posited by the statistical multifractal equations of Lovejoy and Schertzer [7], in the case of temperature behaving like a passive scalar in the horizontal. We will see, however, that T does not always so behave in the vertical. Further evidence of statistical multifractal behaviour can be seen in **Figure 2**, where the probability distributions are shown for millions of points from ER-2 flights taken at 1 Hz – corresponding to 200 m horizontal resolution - during Arctic summer and winter conditions. The fat tails are characteristic; Gaussians are not seen. The Lévy exponent α expresses this, being on average 1.6 whereas a Gaussian has the value 2; its predicted range is $1.5 < \alpha < 2$, see [12, 13]. The implication of this is that the variance of T does not converge, although the mean does.

The evidence for correlations between ozone photodissociation rate with temperature and its intermittency is exemplified in **Figures 5** and **6**.

Figure 3 shows the photodissociation rate $J[O_3]$ plotted against the intermittency of temperature $C_1(T)$ for flights of the ER-2 in the Arctic lower stratosphere for the spring–summer-autumn of 1997 and the winter of 2000. **Figure 4** shows T itself plotted against $C_1(T)$ for the same data. We examine in **Figures 5** and **6** evidence from a particular flight, 19970509 (yyyymmdd), which crossed the terminator in a region of low wind speeds, enabling flight in 'the same air mass' in sunlit and dark conditions. The separate sunlit and dark legs were not long enough for a separate statistical multifractal analysis. The flight was 35 days from summer solstice and had a higher sun than most of the flights on **Figures 3** and **4**. In **Figures 3** and **4**, the intermittency of temperature never drops to zero even when the ozone photodissociation rate does, although approaching it in the coldest and darkest points in winter; see the points in the lower left corners. The intermittency of temperature

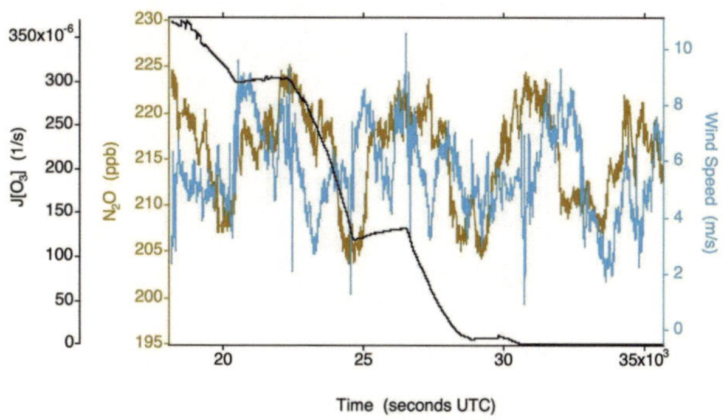

Figure 5.
Observations from the ER-2 on 19970509 (yyyymmdd), when 'racetrack' segments were flown either side of the terminator in a slow moving airmass at about 55 mbar. Black curve, J[O_3]; O_3, blue; T, *red; green, east longitude. Note that temperature is cooler in the dark, while ozone does not change. Like wind speed and the tracer nitrous oxide in **Figure 6**, it is approximately symmetrical about the terminator, where J[O_3] $\approx$ 0.* T *has increased by about 0.4 K in two hours in the sunlit air.*

is highest in the sunniest and warmest points, clustered in the upper right corners of **Figures 3** and **4**. **Figures 5** and **6** can cast some light on the foregoing results, allowing a direct estimate of the radiative heating rate.

In **Figures 5** and **6**, the black trace defines $J[O_3]$ and hence goes to near zero at the terminator and beyond into night. Across the terminator, the low wind speed, ozone and tracer nitrous oxide, while varying about a mean in sunlight and dark, remain constant on average. That is not true of temperature, which is on average 0.4 K higher in sunlit conditions, 'in the same air'. A heating rate of 0.2 K/hour at 55 mbar can be calculated, and successfully checked by computation of the energy influx from radiation and the specific heat of air. The heating rate is consistent with the observation of $J[O_3]$.

In May, and September, the temperature in the sunlit part of the flight legs was on the warm side of the probability distribution compared to the dark side, again 'in the same air'. This result can be seen in **Figures 7** and **8**.

These considerations are also consistent with the concept of heating by unthermalized translationally hot oxygen atoms causing intermittency of temperature.

The behaviour of temperature in the vertical is not that of a well-mixed passive scalar ('tracer' in atmospheric usage). Its scaling is dominated by gravity [12, 13, 21–23]. Experimental evidence is shown in **Figures 9** and **10**.

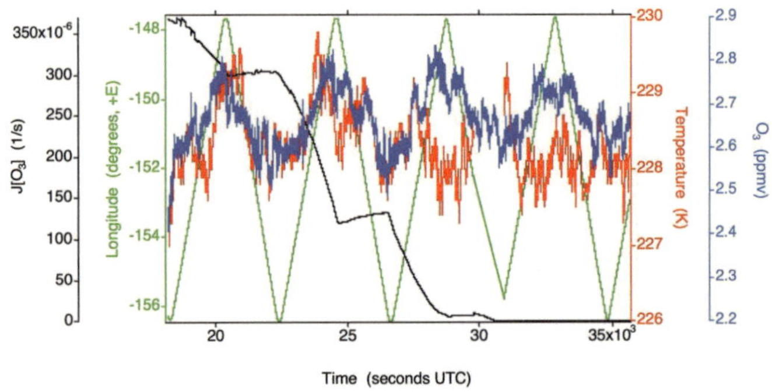

Figure 6.
Same flight as in ***Figure*** *5. Windspeed, light blue; nitrous oxide, brown. In a slow-moving air mass (no more than 3% movement during the flight), wind speed and tracer are approximately symmetrical about the terminator, with temperature in* ***Figure*** *5 being the only variable showing asymmetry about* $J[O_3] \approx 0$.

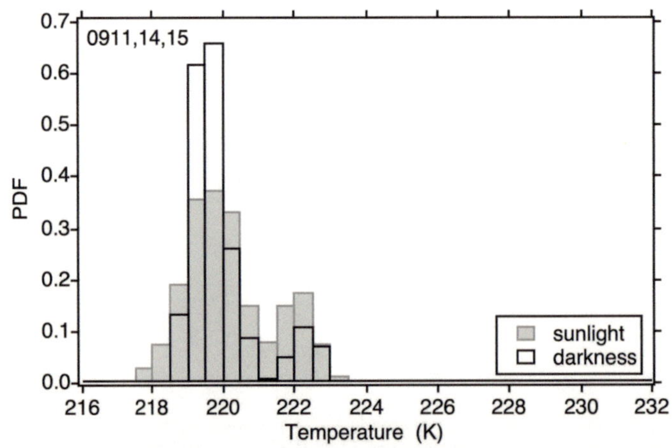

Figure 7.
Probability distribution for temperature, normalised to unity, for the data in ***Figure*** *5. Population has moved from the more probable values on the dark side of the terminator to the sunlit side.*

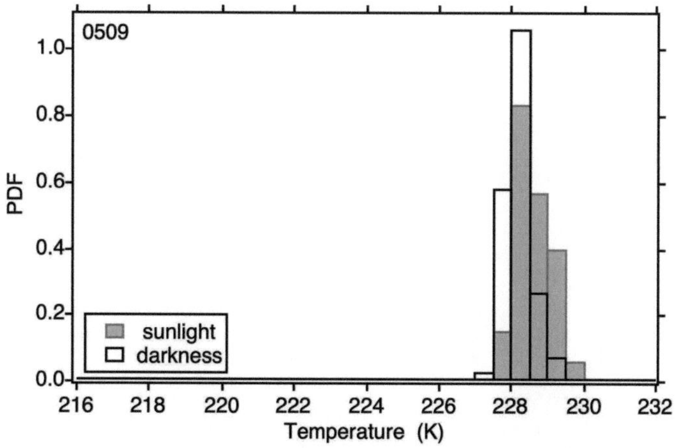

Figure 8.
Probability distribution for temperature, normalised to unity, either side of the terminator for three flights from (65°N,148°W) 19970911 (yyyymmdd), 19970914 (yyyymmdd) and 19970915 (yyyymmdd). As for 20040305 (yyyymmdd), the sunlit data have gained population from the dark data.

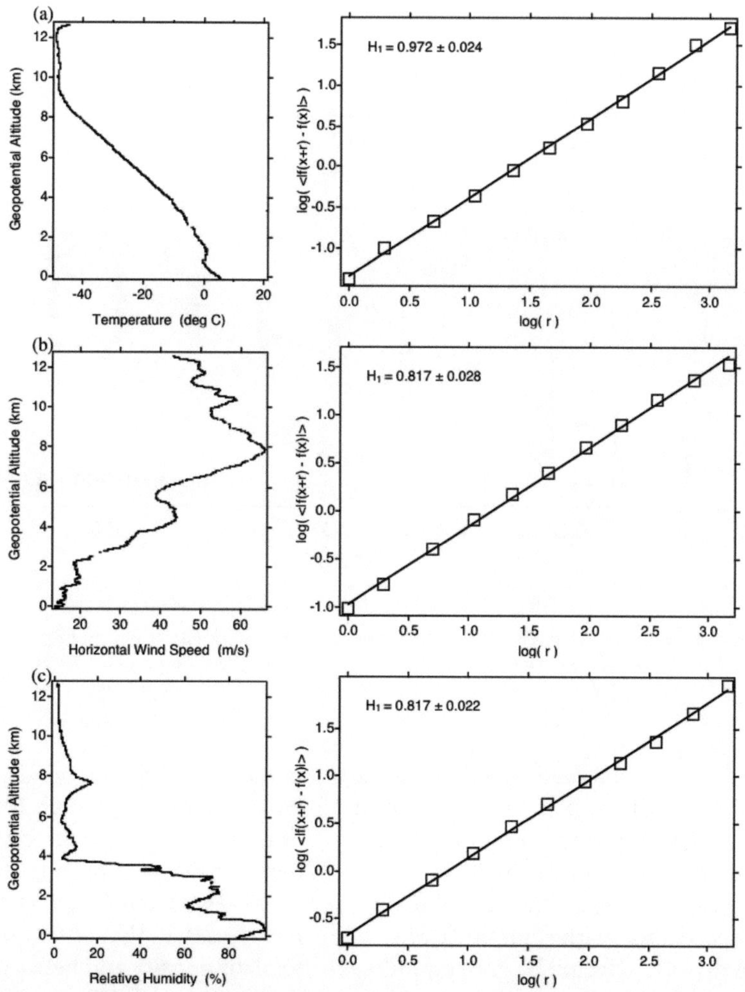

Figure 9.
Observations and H_1 scaling for dropsonde descent at (42° 42′ 56" N, 170° 55′ 30" W), temperature, horizontal wind speed and relative humidity, 20040305.

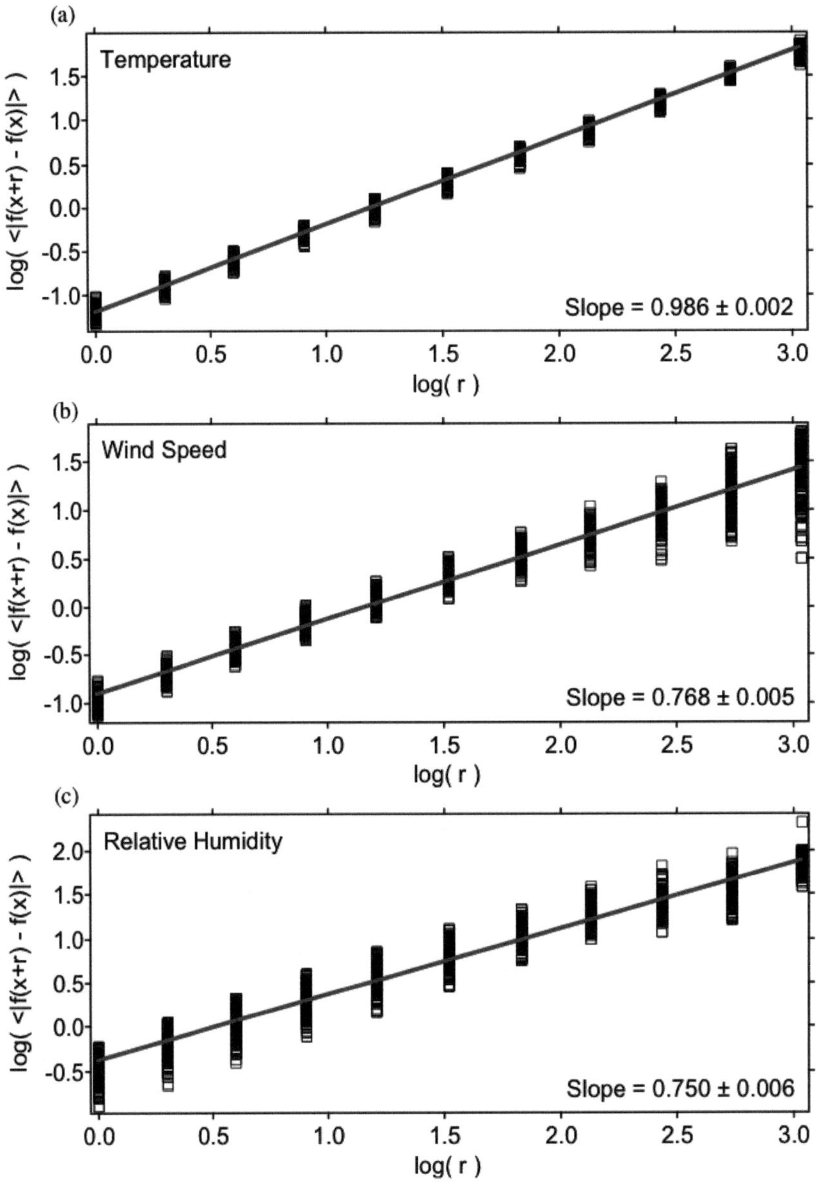

Figure 10.
*Composite variogram for all 885 useable dropsonde descents over the eastern Pacific Ocean, from 15°N to 60°N, like that in **Figure 9** during winter storms 2004, 2005 and 2006. The scaling is excellent for all three of temperature, horizontal wind speed and relative humidity. See [23] for further discussion.*

Figure 9 shows the vertical scaling of the temperature, horizontal wind speed and relative humidity for an individual dropsonde descent in early March 2004 [23]. The scaling exponent of temperature $H_1(T)$ is near unity; it does however have significant intermittency of ≈0.20 (not shown), demonstrating departures from hydrostatic equilibrium. The vertical scaling exponents $H_1(s)$ and $H_1(\mathrm{rh})$ of horizontal wind speed and relative humidity are significantly less than that of temperature, being not directly affected by gravity, unlike the total air density and hence temperature. **Figure 10** shows the grand average composite variogram of 885 dropsondes from the three years of Winter Storms 2004, 2005 and 2006. The vertical scaling of temperature, wind speed and relative humidity is further discussed in [23].

5. Gibbs free energy, irreversibility, the atmosphere and climate heating

Gibbs free energy – 'exergy' – is what enables, via dilution of energy density, the circulation of the atmosphere. It forces directional evolution through the most energetic high-speed molecules, giving the arrow of time to the atmosphere, and its associated irreversibility. Its 2nd Law cost is paid for by the dissipation enabled by the molecules with the most probable speeds easily exchanging energy to maintain an operationally observed temperature while enabling the evolution of vorticity and atmospheric flow, expressed as scale invariant turbulence. The basis for this is discussed in Chapter 7 of Tuck [12], arising from Alder and Wainwright [11]. The energy source is the solar beam, the dissipation is infrared dissipation to space from the infrared active molecules, mainly but not exclusively in the 7–14 μm wavelength window region, by water vapour, water vapour dimers, carbon dioxide, ozone, methane, nitrous oxide and halocarbons. These radiatively active molecules display scale invariance, ensuring that the absorption and emission of radiation operates on all scales, from the mean free path up to 15 orders of magnitude to a great circle. The long-understood fact that atmospheric state and evolution is governed by the interaction between radiation, chemistry and dynamics is extended to the smallest, microscopic scales. The transition from meso to macro scales maintains an operational temperature, and points to the reason the atmosphere 'integrates' a portion of the variability in its constituents and dynamical quantities [14].

Because the change in Gibbs free energy ΔG is equal to or is less than the difference between the change in enthalpy H and the product of the temperature T change and the entropy S change,

$$\Delta G \leq \Delta H - T\Delta S \tag{6}$$

a thermodynamic profit in the case of amelioration of climate heating by fossil fuel burning is not possible by using power generated by those means. Consider that typical coal burning power stations operate at about 35% efficiency; 65% is dissipated. It follows that the entropy cost of any remedial intervention must be paid for by energy derived from renewable sources rather than from fossil fuel combustion. For example, the entropy cost of removing 25% of 400 ppm mole fraction of carbon dioxide from the air will be very high. **Figure 11** shows the behaviour of the Gibbs free energy derived from scaling analysis of the temperature in **Figure 1**, see [14].

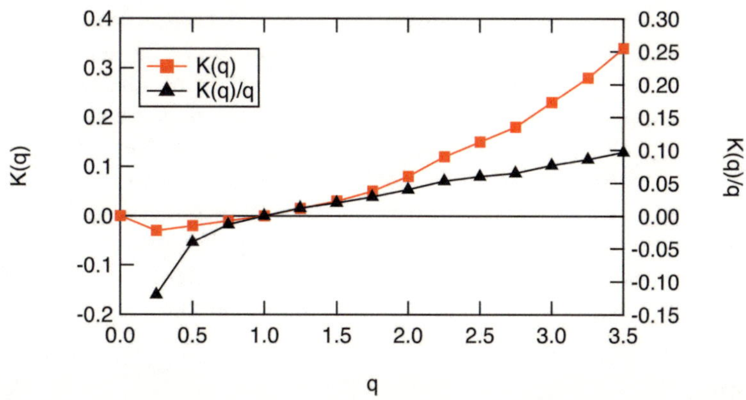

Figure 11.
The behaviour of K(q) *and* K(q)/q *as a function of* q*; see* ***Table 1*** *and Eqs. (1)-(5) for definitions. At* q = 1 *both functions are at or near zero, indicting a steady state condition. There is further discussion in Section 4 of [14].*

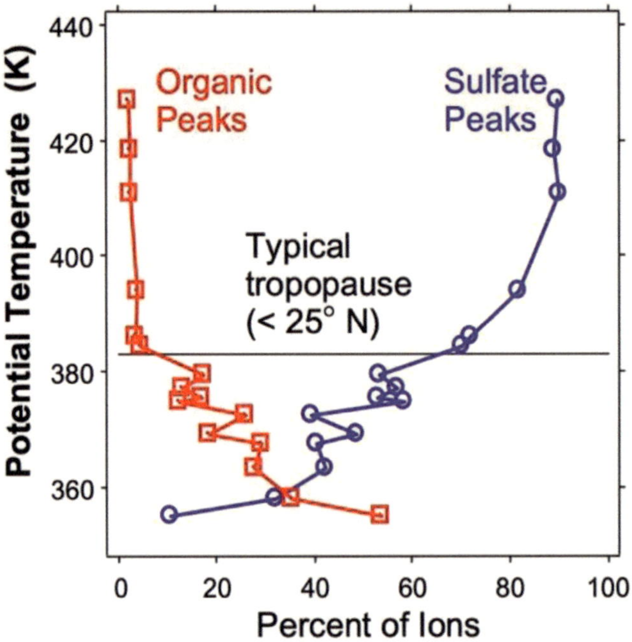

Figure 12.
The relative percentages of organic and sulfate ions in the single particle laser mass spectrometry from the WB57F aircraft [24]. There is significant presence of organic molecules well into the stratosphere. See also [18, 19, 25].

Both Gibbs free energy $K(q)/q$ and the exponent in the partition function $-K(q)$ go through zero at $q = 1$, indicating existence at or near a steady state. Input energy will move the system from steady state towards higher temperatures, with scaling Gibbs free energy enabling movement to more energetic states as $K(q)/q$ becomes more negative. Cooling on the other hand will move the system to less energetic steady states at higher q.

The above considerations apply to geoengineering projects. Further considerations apply in the case of so-called "solar radiation management", such as the limitations of our current understanding of the chemical, radiative and cloud physical roles of aerosols in the lower stratosphere [5, 12, 18, 19]. Aerosols have been shown, over a population of millions of individual particles, to contain 45 different elements from the 92 in the periodic table, not all of course in any single particles. The aerosols are neither internally nor externally mixed at altitudes 5 to 19 km [24].

The organic content of lower stratospheric aerosols has significant effects on radiative transfer [18] and is photochemically influenced [19, 25] (see **Figure 12**).

The advantage of the statistical multifractal approach embodied in Eqs. (1)-(5) is that it can be computed for the system as a whole with adequate observations of wind speed, temperature and if necessary the radiative constituents to use Eq. (6), without having to measure all the chemical constituents of air to make what are in any case inadequate equilibrium quantum statistical thermodynamic calculations. It could also be applied to individual aerosol particles [5] via high-resolution observations and molecular dynamics calculations. These arguments, and those in the preceding sections, suggest a programme of future research.

6. Conclusion

The above arguments show that the changes from fossil-fuel induced heating are irreversible in the strict sense of quantum statistical thermodynamics. This however is expected because the atmosphere is not an isolated system at equilibrium; rather it is an open system far from equilibrium. We show above that it is irreversible by statistical multifractal analysis; **Figure 11** is an example of how it works, showing that to move from an existing steady state will take energy to do work that must be enabled by the dissipation mechanism, which is infrared radiation to space. Conditions to achieve a thermodynamic profit therefore indicate that any remedial action must use renewable energy, which in principle is available in abundance given that the entire biosphere uses only about 1% or so of the incident solar beam. The many uncertainties in climate modelling; in the cloud physical, chemical, dynamical and radiative uncertainties associated with aerosols; in the inability to forecast beyond macroweather time scales all imply that geoengineering, particularly in the form of "solar radiation management", is an extremely uncertain gamble [19]. While the situation is in principle irreversible as the entropy-carrying infrared photons recede into space at the speed of light over the whole 4π solid angle, the simplest and least risky course of action is to use renewably generated Gibbs free energy to move to a sustainable steady state, by reducing and then eliminating fossil fuel burning, and to do it with despatch.

Acknowledgements

Interaction and cooperation with Aiden Hovde, Shaun Lovejoy, Daniel Schertzer and Veronica Vaida have been essential in developing the case presented here.

Author details

Adrian F. Tuck
Atmospheric Physics, Imperial College London, UK

*Address all correspondence to: adrianftuck@gmail.com

References

[1] Donaldson DJ, Vaida V. The Influence of Organic Films at the Air-Aqueous Boundary on Atmospheric Processes. Chemical Reviews 2006; 106(4) 1445-1461.

[2] Yan X, Bain RM, Cooks, RG. Organic Reactions in Microdroplets. Angewandte Chemie International Edition 2016; 55 12960-12972.

[3] Nam I, Lee JK, Nam HG, Zare RN. Abiotic Production of Sugar Phosphates and Uridine Ribonucleoside in Aqueous Droplets. Proceeding of the National Academy of Sciences of the United States of America 2017; 114 12396-12400.

[4] Zare RN. Chemical and Engineering News 2018. Available on line: https://www.cen.acs.org/sponsored-content/what-will-be-chemistrys-next-big-thing.html (accessed 27 May 2018).

[5] Tuck AF. Gibbs Free Energy and Reaction Rate Acceleration in and on Microdroplets, Entropy 2019; 21 1044.

[6] Wilson KR, Prophet AM, Rovelli G, Willis MD, Rapf RJ, Jacobs MI. A Kinetic Description of how Interfaces Accelerate Reactions in Micro-compartments. Chemical Science 2020; 11 8533-8545.

[7] Lovejoy S, Schertzer D. The Weather and Climate: Emergent Laws and Multifractal Cascades. Box 5.1 pp 127-128. Cambridge: Cambridge University Press; 2013.

[8] Schertzer D, Lovejoy S. Generalised Scale Invariance in Turbulent Phenomena. Physico-Chemical Hydrodynamics Journal 1985; 6 623-635.

[9] Schertzer D, Lovejoy S. Physical Modeling and Analysis of Rain and Clouds by Anisotropic Scaling of Multiplicative Processes. Journal of Geophysical Research 1987; 92 9693-9714.

[10] Lovejoy S. Weather, Macroweather, and the Climate. Oxford: Oxford University Press; 2020.

[11] Alder BJ, Wainwright TE. Decay of the Velocity Autocorrelation Function. Physical Review A: Atomic, Molecular and Optical Physics 1970; 1 18-21.

[12] Tuck AF. Atmospheric Turbulence: a Molecular Dynamics Perspective. Oxford: Oxford University Press; 2008.

[13] Tuck AF. From Molecules to Meteorology via Turbulent Scale Invariance. Quarterly Journal of the Royal Meteorological Society 2010; 136 1125-1144.

[14] Tuck AF. Proposed Empirical Entropy and Gibbs Energy Based on Observations of Scale Invariance in Open Nonequilibrium Systems. Journal of Physical Chemistry A 2017; 121 6620-6629.

[15] Goede APH, Tuck A, Burrows JP, Fischer H, Leisner T, Beekmann M, Flaud J-M, Paretzke H, Suppan P, Baier F, Hamacher T, Platt U, Zanis P. Energy and Environment: The Intimate Link. Verhandlungen der Deutschen Physikalischen Gesellschaft, Frühjahrstagung 2008; Umweltphysik Reihe VI Band 43, 311.

[16] Lovejoy S, Schertzer D, Stanway JD. Direct Evidence of Multifractal Cascades from Planetary Scales Down to 1 km. Physical Review Letters 2001; 86 5200-5203.

[17] Tuck AF, Hovde SJ, Richard EC, Gao R-S, Bui TP, Swartz WH, Lloyd SA. Molecular Velocity Distributions and Generalized Scale Invariance in the Turbulent Atmosphere. Faraday Discussions 2005; 130 180-193.

[18] Yu P, Murphy DM, Portmann RW, Toon OB, Froyd KD, Rollins RW, Gao R-S, Rosenlof KH. Radiative Forcing from Anthropogenic Sulfur and Organic Emissions Reaching the Stratosphere. Geophysical Research Letters 2016; 43 9361-9367.

[19] Tuck AF, Donaldson DJ, Hitchman MH, Richard EC, Tervahattu H, Vaida V, Wilson JC. On Geoengineering with Sulphate Aerosols in the Upper Tropical Troposphere and Lower Stratosphere. Climatic Change 2008; 90 315-331.

[20] Duxbury G, Tuck AF, Platt U, Heard DE, Ashfold MNR, Plane JMC, Herrmann H, Jones RL, Smith IWM, Taatjes CA. General Discussion. Faraday Discussions 2005; 130 248-252.

[21] Lovejoy S, Tuck AF, Hovde SJ, Schertzer D. Do Stable Atmospheric Layers exist? Geophysical Research Letters 2008; 35 L01802.

[22] Tuck AF, Hovde SJ. Fractal Behavior of Ozone, Wind and Temperature in the Lower Stratosphere. Geophysical Research Letters 1999; 26 1271-1274.

[23] Hovde SJ, Tuck AF, Lovejoy S, Schertzer, D. Vertical Scaling of Temperature, Wind and Humidity Fluctuations: Dropsondes from 13 km to the Surface of the Pacific Ocean. International Journal of Remote Sensing 2011; 5891-5918.

[24] Murphy DM, Thomson DS, Mahoney MJ. In situ measurements of organics, meteoritic material, mercury and other elements in aerosols from 5 to 19 kilometers. Science 1998; 282 1664-1669.

[25] Ellison GB, Tuck AF, Vaida V. Atmospheric Processing of Organic Aerosols. Journal of Geophysical Research 1999; 104 11633-11641.

Chapter 6

Entropy Based Biological Sequence Study

Bimal Kumar Sarkar

Abstract

SARS-CoV-2 virus strains are taken into consideration for the analysis of digitized sequences of information by means of the notions of entropy. The occurrence of a particular pattern in the corona viral sequence is paid a special attention. The incidence of genetic word is represented in a density means. The incidence frequency of the q-gram genetic word is determined with the help of finite impulse response (FIR) filter along the sequence. It is in turn, used for the determination of the probability distribution of the genetic word incidence as the input for the calculation of entropy in the sequence. The sequence entropy is further used for principal component analysis (PCA) to determine the similarity/dissimilarity between the viral sequences. We have considered seven human corona virus sequences. Entropy based similarity study for SARS-CoV-2 strains is presented in this work.

Keywords: sequences, genetic information, FIR filter, entropy, PCA, corona virus

1. Introduction

The entropy of amino acid sequences in DNA of an organism can be considered as the measure of diversity of proteins. The higher the value of entropy, the greater the possibility of variation in the information content coded by the nucleic acid [1]. This theory is utilized in the present study to understand the variation in the genetic sequences of different novel corona viruses that have infected people across the world leading to one of the world's biggest pandemics. The pandemic itself highlights the importance of tracking the dynamics of viral transmission in real-time. Moreover, as the virus mutates frequently, each sequence is studied and compared with others to understand the variation of information that is transmitted from one species to the other. Hyper-variable genomic hotspot for the novel coronavirus SARS-CoV-2 has already been identified by Wen et al. [2]. Likewise, the similarities in the genetic code would also provide important information in understanding the virus and its prevention.

Corona virus molecule has a single-stranded, positive-sense RNA genome of length of approximately 27 to 32 kilobases (kb). The genome sizes of HCoV-229E and HCoV-NL63 are approximately 27.5 kb, and it is more than 30 kb for HCoV-OC43 and HCoV-HKU1. It possesses the RNA harbors a 50-cap structure and a 30-polyadenylate tail which enable to play a role of messenger RNA (mRNA) [3–10].

This study presents identification and analysis of regions of similarity in SARS-CoV genetic sequence [11–13]. According to information theory, individuality of a species can be aggregates that propagate information from past to future. The Shannon Entropy is considered as a measure for the order/disorder state of nucleotide sequences of the DNA [14]. The information in a genetic code is comprised of an alphabetic sequence of the four letters A, C, G, and T, which symbolizes the four nucleotides, namely, adenine (A), cytosine (C), guanine (G) and thymine (T). The sequences have been recognized for most of the SARS-CoV-2 genes and are accessible in computer readable form. The probability of occurrence of a combination of a group of symbols in a sequence is the measure of order in a sequence. An alignment free approach of DNA sequence analysis, *n*-mer/word frequency estimation, is attempted in this work.

2. Methodology

Our method is based on the observation through a sliding "counter "of width W over DNA sequence [15]. A certain number of q-grams called as bins are set in the counter. As there are only four letters in the DNA alphabet, viz., {A, C, G, T} the number of all combinations of q-grams in a DNA sequence is 4^q.

Definition 1. q-gram of Sequence.

Given a sequence 'seq', when a window of length q slides over the characters of 'seq', its q-grams are formed. For a sequence 'seq, there are $|seq| - (q - 1)$ *q-grams.*

The number of all possible q-grams or called as "bin" is 4^q. Bins can be arranged in lexicographic order, and b_i is used to denote the ith bin in this order. All the possible bins are denoted as:

$$B_q = \{b_1,\ b_2,\ \dots\dots\ b_{4^q}\} \tag{1}$$

Example 1. One-gram bins are $B_1 = \{A,\ C, G, T\}$, consisting 4 bins. Two-gram bins are $B_2 = \{$AA, AC, AG, AT, CA, CC, CG, CT, GA, GC, GG, GT, TA, TC, TG, TT$\}$, consisting 16 bins.

Definition 2. Bin Signature.

For a sequence, the q-gram bin signature, S_j is a mapping with the bin b_j $(b_j \in B_q)$ where i^{th} bit in S_j, is corresponding to the presence or absence of b_j. For a sequence 'seq, there are $|seq| - (|b_j| - 1)$ bits in S_j.

Example 2. Consider a sequence, S = "AACTCG". Its two-grams (q = 2) signature in the sequence is S_2 = [0 1 0 0 0].

Definition 3. Filter.

A sequence x[n] is filtered through mapping of the sequence into output sequence y[n] via a weighted window b by means of the convolution summation as

$$y[n] = \sum_{i=0}^{k} b_i x[n - i] \tag{2}$$

b is independent of x[n] and y[n], where n is the time index. y[n] is the response of the filter to input signal x[n]. The filter is finite impulse response (FIR) digital filter. The term digital filter arises because it operates on discrete-time signals. Finite impulse response arises because the filter output is computed as a weighted, finite term sum, of past and present (**Figure 1**).

Example 3: Weighted filter output of S_A with the weighted window β = [0.2 0.1 0.3 0.4] is as follows:

S_A = [1 1 0 0 0 0].

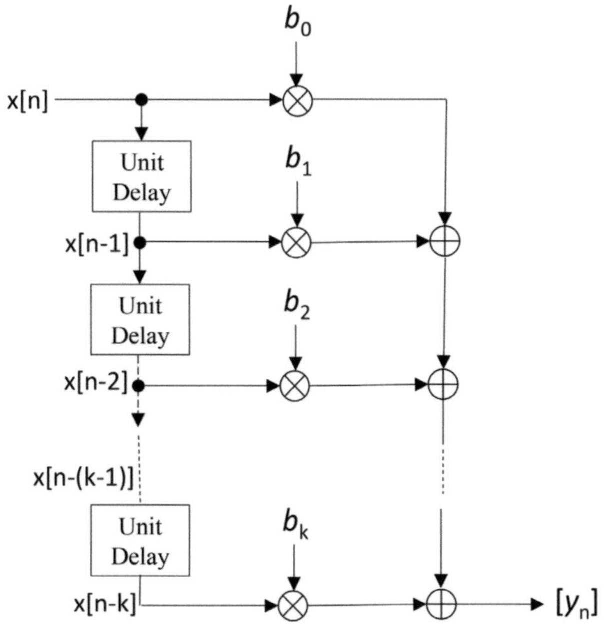

Figure 1.
Block diagram of finite impulse response (FIR) digital filter.

$$y_A[n] = \sum_{0}^{3} \beta_k S_A[n-k] \text{ with } \beta_0 = 0.2, \beta_1 = 0.1, \beta_2 = 0.3, \beta_3 = 0.4.$$

$$\Rightarrow y_A[n] = \beta_0 S_A[n] + \beta_1 S_A[n-1] + \beta_2 S_A[n-2] + \beta_3 S_A[n-3]$$

y_A = [0.2 0.3 0.4 0.7 0.4 0]; Similarly for other nucleotide viz., C, G, T, the output is obtained as,

y_C = [0.2 0.0 0.2 0.1 0.5 0.5]; y_G = [0.0 0.0 0.0 0.0 0.0 0.2]; y_T = [0.0 0.0 0.0 0.2 0.1 0.3];

For nucleotide density calculation, evenly distributed window of unit value is considered. As explained, the output of the convolution summation represents the nucleotide density along the sequence. The detail algorithms for bin construction, bin signature, filter operation is displayed in **Tables 1–3** respectively.

Input: q - length of bin **Output:** set of bins $B_q = \{b_1, b_2, \ldots\ldots b_{4^q}\}$

```
0 ← bincount;
4^q ← n;
cell(1,n) ← bin;
for first = 1:4 do
    for qth = 1:4 do
    convert integer to nucleotide character ([first ... qth]) ← binq;
    bincount = bincount +1;
    binq← bin{bincount};
    end
end
bin ← B_q
```

Table 1.
Bin construction.

Input: Sequence (seq), bin (b) **Output:** Bin Signature

```
m ← length (seq);
nbin← length (b);
for i ← 1….m - (nbin - 1) do
   if seq (i: i + nbin-1) = b then
      signature (i) = 1
   else
      signature (i) = 0
end
signature ← Bin Signature
```

Table 2.
Bin signature.

Input: *BinSignature*, *window* **Output:** *filter*

```
w ← length (window);
window = 1/w*array of ones(1,w);
0 ← sum
for i ← 1…. length (window) do
make array of zeros with length of i − 1 ← zero
sum = sum + window (i) * array[zeros BinSignature(1:(length(BinSignature)-(i-1)))]
end
filter ← sum
```

Table 3.
Filter.

3. Sequence analysis

The filter output is taken as a density distribution for DNA sequences. The density distribution is based on q-gram word density, which in turn is considered for the determination of Shannon Entropy as

$$y_i = -\sum_{j=1}^{q} p_{ij} \log p_{ij} \tag{3}$$

where p_{ij} is the probability of appearance of the jth genetic letter at ith position in the genetic sequence. Further we want to find a similarity/dissimilarity measure between two entropy distributions $\rho_i = (y_{i1}, y_{i2}, \ldots, y_{in})$ and $\rho_j = (y_{j1}, y_{j2}, \ldots, y_{jn})$. We construct the data matrix D comprising elements $[\rho_1, \rho_2, \ldots, \rho_m]'$, where m is the number of sequences. Principal Component Analyses (PCA) is used to estimate scores between density distributions such that it reduces multidimensional data sets to lower dimensions with the consistent original data matrix [16].

We determine the dissimilarity between two sequences from the scores in the first three principal components by computing the Euclidean distance between pairs of density distributions in the m-by-n data matrix D. Rows of D correspond to sequence (observations) and columns correspond to position index in the sequence (variables). Thus, Euclidean distance X is a row vector of length $m\ (m-1)/2$, corresponding to pairs of observations in D. The distances are arranged in the order $(2, 1), (3, 1), \ldots, (m, 1), (3, 2), \ldots, (m, 2), \ldots, (m, m-1)$. X is used as a dissimilarity matrix in clustering or multidimensional scaling. An unweighted pair group method with arithmetic mean (UPGMA) is employed on PC scores for the construction of a phylogenetic tree [17]. UPGMA uses a local objective function to construct a rooted bifurcating tree.

4. Results and discussions

The nucleotide density distribution was obtained through FIR filter. We have calculated the density distribution for one-, two-, three-, gram nucleotide for different species. Secondly we have calculated entropy distributions ρi = (yi1, yi2, ... , yin) and ρj = (yj1, yj2, ... , yjn). The variation of entropy with position for all other sequences are calculated for the above three combinations. The entropy values were found to be minimum for mono-mer density distributions in individual sequences while increasing linearly for di-mers and codons respectively. Observations based on the position of the n-mers in sequences of SARS-CoV-2 DNA reveals significant minimum entropy regions for codons. **Figure 2** shows the entropy profile calculated over 29000 bases for 7 DNA sequences. Similar analysis profile for mono-mers and di-mers does not show overlapping regions for different sequences. This suggests that codons are more effective in transferring information through different species. Codon bias has been reported for HIV 1 virus [18]. Therefore, it can be inferred that in various novel coronavirus strains, the codons at specific positions are the highest bias representing minimum entropy and hence carry the maximum information. Further studies with the sequences of these loci can be useful genetic engineering for developing vaccines or taking control over the spread of the second wave of the pandemic.

We have chosen seven SARS Corona virus sequences (SARS-COV) from various countries. The details of the organism are presented in **Table 4**.

Based on FIR filtering, firstly the nucleotide density distribution is generated. We have calculated the density distribution for one-, two-, three-, gram nucleotide for different species. Secondly we have calculated entropy distributions ρi = (yi1, yi2, ... , yin) and ρj = (yj1, yj2, ... , yjn). **Figure 3** displays the spatial variation of the entropy along the SARS-COV sequence for seven species.

In fact it is inconvenient to realize all the entropy variation in 2D graphical representation. For example, the organism HKU1 shows the positions where it possesses the minima in entropy values. Some are demonstrated at the positions, around 7400, 10000, 23000 etc. the Amsterdam strain, NL63 has shown minima at around 7300, 8000 etc. But other strains exhibit their entropy representation in a crowded manner. It is difficult to understand the variation for them differentially. Rather it is more comprehensive to show the entropy variation for all sequences (total 7) in a single panel. It has been shown in **Figure 3**.

The present work intends to assess the variability and complexity at each nucleotide site with the calculation of entropy for each position using the Shannon entropy formula, Eq. (2). The low entropy regions around 7400 and 9000 position

Sequence No.	Strain	Accession No.	Place
1	Wuhan-Hu-1	MN908947	Wuhan
2	CV7	DQ898174	Canada
3	MERS-CoV /C1272	MH734115	Kenya
4	HCoV-OC43	KU131570	UK/London
5	NL63	DQ445912	Amsterdam
6	HCoV_229E	MN306046	Seattle/USA
7	HKU1	MH940245	Thailand

Table 4.
SARS-COV strains with their complete genome sequence, accession no. and source.

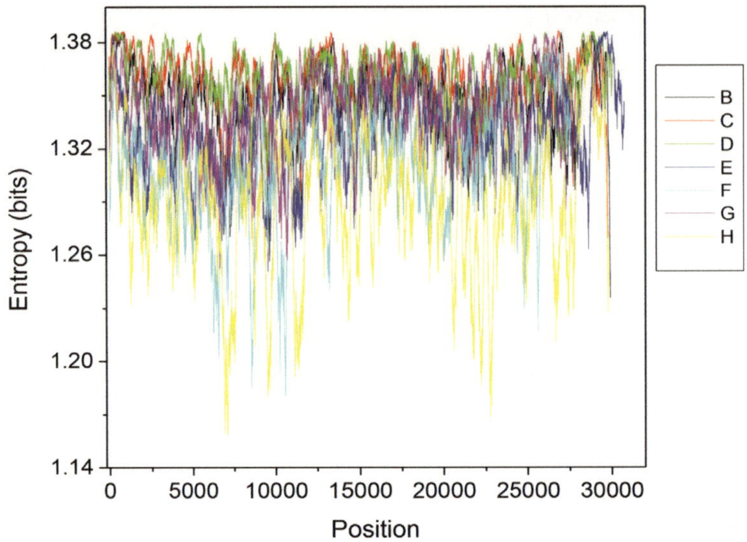

Figure 2.
Entropy profile of seven SARS-COV sequence. Entropy is calculated based on single nucleotide distribution. Sequences are: B: Wuhan-Hu-1; C CV7; D: MERS-CoV/C1272; E: HCoV-OC43; F: NL63; G: HCoV_229E; H: HKU1 (see ***Table 4****).*

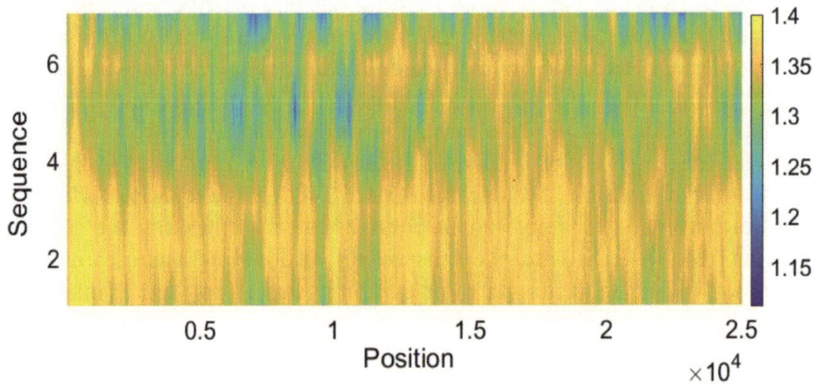

Figure 3.
Entropy profile of 7 SARS-COV sequences. Entropy is calculated based on single nucleotide distribution. Sequences are represented as number starting from 1 to 7. (see ***Table 4****).*

are common to all 7 sequences (**Figure 3**). Entropy (Y_i) is an important parameter for the understanding of sequential stability. Y*i* becomes maximal when all symbols occur at equal probability. On the other hand, Y*i* becomes the least if one symbol occurs at probability 1 and in that case the other symbols will be forbidden. It means that lower the value of entropy the site is more stable without much complexity. Under this assumption, the zone around the site 7400 and 9000 position are most stable for all strain/species. It may find a good structural relationship between the regions of low entropy and the secondary structure of proteins which include α-helix, β-sheets and loops regions.

Strain no. 4–7 (HCoV-OC43; F: NL63; G: HCoV_229E; H: HKU1) show the stability with lower entropy around 8 K, 9 K, 11 K, 12 K site position. But this behavoiu is not exhibited in case of the strains numbers 1–3 (Wuhan-Hu-1; C CV7; D: MERS-CoV/C1272). If one can go through these strains, as a whole, it is noticed that the entropy is increasing or in turn the complexity is more. It is an indication of

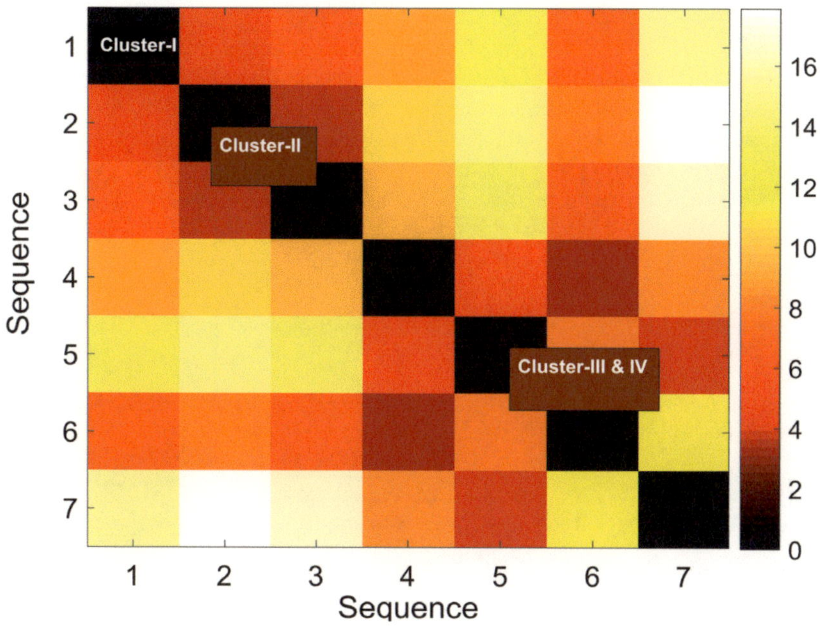

Figure 4.
Dissimilarity matrix for 7 SARS-COV sequences.

evolutionary development among the SARS-COV strains. Based on site entropy we prepared the dissimilarity matrix for the sequences (**Figure 4**).

The dissimilarity matrix demonstrates the existence of 4 different clusters. One can see that the SARS-COV sequences in a cluster shows less dissimilarity among themselves. In other way to mean that the sequences have much similarity residing in a cluster [19]. The COVID sequence appearing in cluster I is typically from Wuhan, China. The Wuhan virus genome sequence examination found β-CoV strain [20]. The Wuhan novel β-CoV revealed 88% similarity with the sequence of two bat-derived SARS-COV, bat-SL-CoVZC45 and it was named "SARS-CoV-2" by the International Virus Classification Commission. The genome of SARS-CoV-2 sequence has the similarity with the typical CoVs. It encompasses more than ten open reading frames (ORFs). The first ORFs covers about two-thirds of viral RNA, which get translated into two large polyproteins, pp1a and pp1ab. These proteins assist to form the viral replicase transcriptase complex [21]. The remaining one-third of viral RNA take part in translation of four structural proteins: spike (S), envelope (E), nucleocapsid (N) and membrane (M) proteins [22].

Cluster-II comprising of two strains CV7, MERS-CoV, belong to β-CoV genera, which also includes SARS-CoV-2 strain as placed singly in cluster-I. Two HCoVs of strains HCoV-229E and HCoV-OC43 being placed in the mixed Cluster of III and IV, are the members of α-CoV genera. From the cluster presentation (**Figure 5**), it will be understood that they belong to cluster-III. Remaining two strains, NL63 and HKU1 are placed in cluster IV.

Phylogenic relation among the strains is represented in **Figure 6**. We obtain the phylogenetic tree of the data set based on unweighted pair group method with arithmetic mean (UPGMA) on PC scores. Phylogenetic tree analysis clearly shows the relationship among all COVID strains under each cluster. We further sub-cluster in each cluster based on their genetic distance (GD). We have considered PC score to determine the dissimilarity or genetic distance between two organisms.

Explicitly the COVID strains are placed in a cluster description (**Figure 5**). The scores are determined in the principal component analysis. Three principal

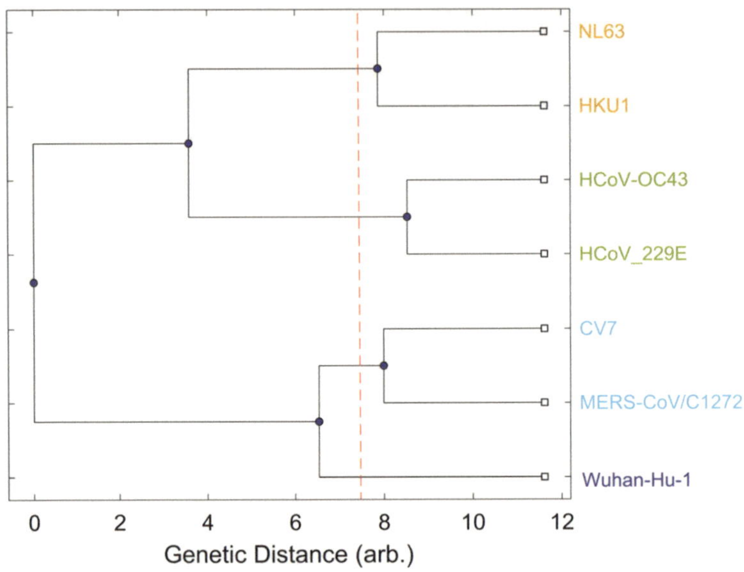

Figure 5.
Scatter plot PC values for 7 SARS-COV sequences: Cluster-I (Wuhan-Hu-1) is encircled with deep blue color). Cluster-II (CV7 and MERS-CoV) is encircled with light blue color. Cluster-III (HCoV-229E and HCoV-OC43) is encircled with green color. Cluster-IV (NL63 and HKU1) is encircled with yellow color.

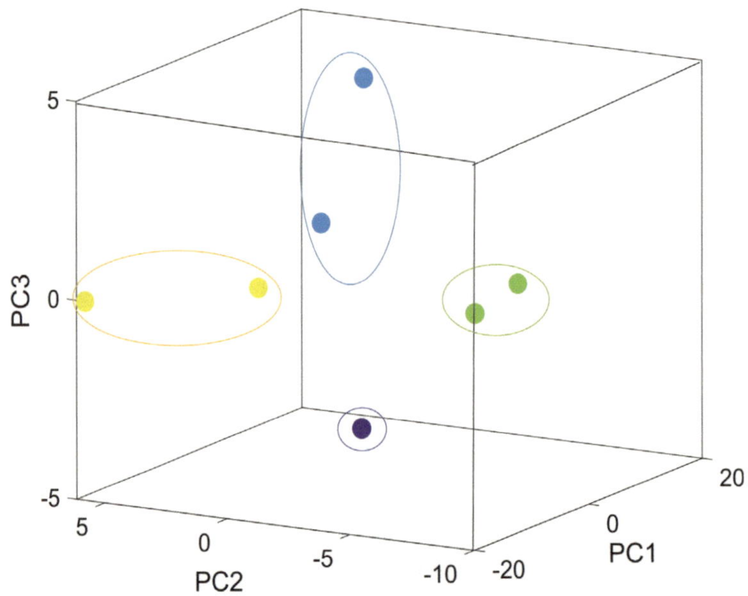

Figure 6.
The phylogenetic tree of 7 SARS-COV sequences.

components are taken into consideration. Each strain is represented as state point by scatter plot in the three PC space. Cluster presentation is well agreement with phylogenic relations. Wuhan-Hu-1 strain is well isolated from all other strains. It belongs to cluster-I. Each of other three clusters possess two-member strain. Cluster-II comprises of two strains CV7 and MERS-CoV belonging to β-CoV genera (encircled with blue color ellipse in **Figure 5**). Already it is mentioned in the previous section that the strains HCoV-229E and HCoV-OC43 exist in Cluster of III.

It is displayed by two state points encircled in green colored ellipse. Remaining pair of strains, NL63 and HKU1 are placed in cluster IV which is marked by yellow colored ellipse.

5. Conclusions

The entropy has been used to select SARS-COV genome regions for stability zone detection. Even though a great deal of genetic variation is generally found, the present entropy calculation is sufficient to observe low informational complexity regions, which are representation of the conserved sites of the sequence. The low entropy regions are related to important functional domains in the proteins of these viruses. Based on entropy calculations seven SARS-COV genomes have been phylogenically described. The clusters of the genome formation is well understood.

Acknowledgements

The authors are thankful to the Department of Physics, Adamas University for providing computational facility. The authors acknowledge the collection of sequence data from NCBI gene bank.

Author details

Bimal Kumar Sarkar
Department of Physics, Adamas University, Kolkata, India

*Address all correspondence to: bks@physics.org.in

References

[1] Hasegawa, M. and Yano, T.A., The genetic code and the entropy of protein. Mathematical Biosciences, 24(1–2), 169–182 (1975).

[2] Wen, F., Yu, H., Guo, J., Li, Y., Luo, K. and Huang, S., Identification of the hyper-variable genomic hotspot for the novel coronavirus SARS-CoV-2. The Journal of Infection, 80 (60), 671–693 (2020).

[3] P. Zhou, X.L. Yang, X.G. Wang, et al., A pneumonia outbreak associated with a new coronavirus of probable bat origin. Nature, 579, 270–273 (2020).

[4] E. de Wit, N. van Doremalen, D. Falzarano, et al., SARS and MERS: recent insights into emerging coronaviruses. Nat. Rev. Microbiol., 14, 523e534 (2016).

[5] J.T. Wu, K. Leung, G.M. Leung, Nowcasting and forecasting the potential domestic and international spread of the 2019-nCoV outbreak originating in Wuhan, China: a modelling study. Lancet, 395 (10225), 689–697 (2020).

[6] S. Su, G. Wong, W. Shi, et al., Epidemiology, Genetic recombination, and pathogenesis of coronaviruses. Trends Microbiol., 24, 490e502 (2016).

[7] S. Perlman, J. Netland Coronaviruses post-SARS: update on replication and pathogenesis, Nat. Rev. Microbiol., 7, 439e450 (2009).

[8] R. Lu, X. Zhao, J. Li, et al., Genomic characterisation and epidemiology of 2019 novel coronavirus: implications for virus origins and receptor binding. Lancet, 395, (10224) 565–574 (2020).

[9] A.R. Fehr, S. Perlman, Coronaviruses: an overview of their replication and pathogenesis. Methods Mol. Biol,, 1282, 1e23 (2015).

[10] P.S. Masters, The molecular biology of coronaviruses. Adv. Virus Res, 66, 193e292 (2006).

[11] X.Y. Ge, J.L. Li, X.L. Yang, et al., Isolation and characterization of a bat SARS-like coronavirus that uses the ACE2 receptor. Nature, 503, 535e538 (2013).

[12] Chinese SARS Molecular Epidemiology Consortium, Chinese Molecular evolution of the SARS coronavirus during the course of the SARS epidemic in China. Science, 303, 1666e1669 (2004).

[13] V.S. Raj, H. Mou, S.L. Smits, et al., Dipeptidyl peptidase 4 is a functional receptor for the emerging human coronavirus-EMC. Nature, 495, 251e254 (2013).

[14] Schmitt, A.O. and Herzel, H., Estimating the entropy of DNA sequences. Journal of theoretical biology, 188(3), 369–377 (1997).

[15] Saha, P. and Sarkar, B.K., Entropy based analysis of genetic information. Journal of Physics: Conference Series, 1579 (1), 012003 (2020).

[16] Novembre, J., Stephens, M., Interpreting principal component analyses of spatial population genetic variation. Nat Genet., 40, 646–649 (2008).

[17] JF Yu, JH Wang, X. Sun, Analysis of similarities/dissimilarities of DNA sequences based on a novel graphical representation. Commun Math Comput Chem, 63, 493–512 (2010).

[18] Grantham R, Gautier C, Gouy M, Jacobzone M, Mercier R., Codon catalog usage is a genome strategy modulated for gene expressivity. Nucl Acids Res., 9 (1), 213–213 (1981).

[19] Chan J F-W, Yuan S., Kok K-H., Wang K-KChu H., A familial cluster of pneumonia associated with the 2019 novel coronavirus indicating person-to-person transmission: a study of a family cluster. The Lancet, 395 (10223), 514–523 (2020).

[20] Lu R., Zhao X., Li J., Niu P., Yang B., Wu H., Genomic characterisation and epidemiology of 2019 novel coronavirus: implications for virus origins and receptor binding. The Lancet, 395 (10224), 565–574 (2020.

[21] A.R. Fehr, S. Perlman, Coronaviruses: an overview of their replication and pathogenesis. Methods Mol. Biol., 1282, 1e23 (2015).

[22] K. Knoops, M. Kikkert, S.H. Worm, et al., SARS-coronavirus replication is supported by a reticulovesicular network of modified endoplasmic reticulum. PLoS Biol., 6, e226 (2008).

Section 3

Wind Energy

Chapter 7

Aerodynamic Analysis and Performance Prediction of VAWT and HAWT Using CARDAAV and Qblade Computer Codes

Tayeb Brahimi and Ion Paraschivoiu

Abstract

Wind energy researchers have recently invited the scientific community to tackle three significant wind energy challenges to transform wind power into one of the more substantial, low-cost energy sources. The first challenge is to understand the physics behind wind energy resources better. The second challenge is to study and investigate the aerodynamics, structural, and dynamics of large-scale wind turbine machines. The third challenge is to enhance grid integration, network stability, and optimization. This chapter book attempts to tackle the second challenge by detailing the physics and mathematical modeling of wind turbine aerodynamic loads and the performance of horizontal and vertical axis wind turbines (HAWT & VAWT). This work underlines success in the development of the aerodynamic codes CARDAAV and Qbalde, with a focus on Blade Element Method (BEM) for studying the aerodynamic of wind turbines rotor blades, calculating the induced velocity fields, the aerodynamic normal and tangential forces, and the generated power as a function of a tip speed ration including dynamic stall and atmospheric turbulence. The codes have been successfully applied in HAWT and VAWT machines, and results show good agreement compared to experimental data. The strength of the BEM modeling lies in its simplicity and ability to include secondary effects and dynamic stall phenomena and require less computer time than vortex or CFD models. More work is now needed for the simulation of wind farms, the influence of the wake, the atmospheric wind flow, the structure and dynamics of large-scale machines, and the enhancement of energy capture, control, stability, optimization, and reliability.

Keywords: wind turbine, DMS, performance, loads, dynamic stall, induced velocity, power, normal and tangential forces

1. Introduction

Wind energy has been recognized as one of the fastest-growing energy sources in the world. The U.S. Energy Information Administration [1] reported that in 2020 the wind's annual electricity generation exceeds 300 million MWh in the U.S., which surpassed the hydroelectric generation by 26 million MWh. In the last decade, the global cumulative wind power capacity installed has increased from about 200 gigawatts (GW) in 2010 to more than 650 GW in 2019, as shown in **Figure 1** [2, 3]. Compared to 2018, the global wind power installed capacity in 2019 represents a percentage increase of 19% with a 10% increase in new installation, which is the second-largest increase in the last decade. This increase was the result of the largest market from China (237029 MW), EU-28 (192231 MW), the USA (105433 MW), and India (37529) [3, 4]. With the continuous technology development of renewable energy, wind power becomes predominant than hydroelectric, biomass, or geothermal energy [5]. Currently, in Europe alone, electricity generated from wind turbines covers up to 11% of the electricity demand; by 2020, it will increase to 16.5%, and by 2030 it is expected that renewable energy could serve at least 27% of Europe electricity need, and will generate over three million jobs. Globally, in 2020 the anticipated wind energy will be dominated by China (38%), Europe (28%), US (16%), and India (7%) and by 2030, based on central scenario, it is expected to have a cumulative installation of 323 GW in Europe alone. Governments, policymakers, and energy utility companies currently employ a wide array of tools to encourage the deployment of various renewable energy technologies including investments, funds, cash, and tax credits incentives.

Before the COVID-19 pandemic, the Global Wind Report published by the Global Wind Energy Council expected a record of new wind installed capacity of 76.7 GW in 2020 [3]. However, given the unpredictable effects of COVID-19 on various renewable energy sectors, it is expected that the wind energy market will generally be slowing down, as the current control of the virus in the US, Europe, and China is still difficult to predict [2]. The International Energy Agency [6], in its new report on the market update outlook for 2020 and 2021, forecasts a 12% decrease in wind power growth compared to 2019. Statista, the online portal for statistics, reported that the global wind market is expected to add 73 gigawatts only instead of the previous predicted installed capacity of 76.7 [7]. The downturn is primarily attributed to project delays instead of cancelations. Yet, there is still a steady increase in global wind installed capacity; for example, the U.S. added 1821 megawatts (MW) of new installed capacity during 2020 Q1 [1], and India added 0.2 GW during January–March 2020 [8]. The International Energy Agency forecasts that over half of Europe's wind growth will come from the Netherlands, Germany, Sweden, Spain, and the U.K. There are currently up to 205 GW wind power installed capacity in Europe, representing 15% of EU-28 of the electricity consumed in 2019 [4].

With the growing ambition and enthusiasm on using the power of the wind to generate clean energy, added to the increasing investments and the drop in wind turbine pricing, more fundamental research and exploration is needed in the design of such wind turbine machines, including environment, social, and economic aspects of meeting the future functionality of large-scale deployment in both onshore and offshore areas. The present study aims to tackle one of the major problems that the wind turbine research community is concerned with [9, 10] by focusing on the aerodynamic loads and performance of vertical and horizontal performance wind turbines (VAWT & HAWT). The study provides a thorough mathematical and physics modeling of VAWT and HAWT aerodynamics using the Double-Multiple-Stream tube model (DMS) and blade element momentum theory (BEM) [11, 12].

2. Previous work

In the last few years, significant research activities have been devoted to designing large-scale wind turbines with high hub height and large rotor diameter, making wind turbines the world's largest rotating machines. Wind turbine rotor diameters were in the range of 5 to 15 m during the 1980s, with an average capacity of 30 kilowatts. In the early 1990s, the wind turbine installed power reached 500 kW with a rotor diameter of 30 m. In 2000, the wind turbine installed power reached 1.0 MW with a rotor diameter of 50 m [13]. Since then, the size of wind turbines is getting larger and larger; in 2018, the power capacity increased to 2.6 MW with a diameter of 110 m, many wind farms have rotor diameters of up to 120 m 5-MW installed power. Current wind turbine technologies and advances in aerodynamics and structural analysis produce lighter and larger wind turbine machines with increased annual energy production [14]. Modern wind turbines may now reach 164 m of the rotor diameter and a power rating of 9.5 MW, such as the Vestas V164–9.5 MW. According to Statista [15], by 2021, the state-of-the-art wind turbine's rotor diameter is estimated to reach 220 meters. Recently, General Electric revealed the Haliade-X, the largest prototype ever designed by the company for a new offshore wind turbine and the most powerful wind turbine machine operating at a 13 MW power output and a rotor diameter of 220 m [16].

With the growing development of wind turbines on a large scale and the increasing amount of wind electricity generated, the current wind energy market needs to be a more competitive, cost-effective, and reliable renewable energy source. The International Renewable Energy Agency (IRENA) [13] reported several programs, projects, and research been introduced to investigate the development of such wind turbine machines and stimulate their commercial growth. The plans include innovation in rotor blade design and materials, optimization of power electronics, incorporating smart/intelligent wind turbines, and using recycling of materials the vast amount of materials used in the wind energy sector.

A long-term strategy to address the scientific and current wind turbine technology has been initiated by the European Academy of Wind Energy (EAWE) in 2016 [10]. The goal was to study and analyze the main barriers and priorities and promote cooperation among researchers in fundamental and applied sciences of wind turbines as more fundamental research and exploration are needed to design such large wind turbines. The EAWE presented and discussed 11 research challenges in wind energy development, namely the materials and structures, the wind and turbulence, the Aerodynamics of wind turbines, the control and system identification, the electricity conversion, the reliability and uncertainty modeling, the design methods, the hydrodynamics, the soil characteristics and floating turbines, the offshore environmental aspects, the wind energy in the electric power system, and the societal and economic factors of wind energy. Three years later, in the United States, the wind energy researchers [9] from the US Department of Energy at the National Renewable Energy Laboratory (NREL) invited the scientific community to interdisciplinary collaboration to tackle three significant wind energy challenges to transform wind engineering into one of the significant global low-cost power generation sources. The first grand challenge is to improve the understanding of the physics behind the wind resource and atmospheric flow in the critical region where the wind machine operates, the second big challenge is to tackle the corresponding structural and dynamics of large-scale rotating wind turbine machines, and the third grand challenge is the enhancement of energy capture, control, network stability, grid integration, optimization, and reliability. Nevertheless, to stay competitive, the cost of electricity generated from wind turbines must continue to decline.

Up to 75% of the overall costs of wind turbine energy production are attributed to upfront costs. In terms of the wind turbine machine's performance and cost, rotor blades are considered the most significant wind turbine parts. The aerodynamic design and optimum shape of the rotor blades, with a high lift to drag ratio, directly impact the wind turbine performance and power generated. There are currently two types of modern wind turbine design: the Horizontal Axis Wind Turbine (HAWT) as the traditional wind pump, and the Vertical Axis Wind Turbine (VAWT) as the Darrieus design model. In both cases, the wind kinetic energy is extracted by the turbine's blades and transformed into electrical power. Both wind turbine machines are currently used offshore and onshore to generate electricity. Although the HAWT is widely used, the VAWT offers a promising alternative due to its mechanical and structural simplicity of harnessing wind energy, scaling down, safety and accepting wind flow from any direction. However, this simplicity encounters a significant challenge during the simulation and computation of the aerodynamic loads. Indeed, during each rotation, the rotor blades encounter the wake it generated, in addition to the wake generated by the rest of the blades, and operate in a dynamic stall regime [11, 17]. Adding to this is the fluctuating nature of the loads due to wind turbulence affecting the wind turbine's planned service life and the power generated, as reported in [18, 19].

Many aerodynamic models for predicting the forces and the power generated by a wind turbine have been developed. A complete state-of-the-art review, including the appropriate references, is given by [17, 20–25]. The wind turbine loads analysis can be achieved using three effective methods: the momentum method through Blade Element Momentum (BEM), the vortex theory, and the Computational Fluid Dynamics (CFD) method, and more recently using artificial intelligence (AI) to predict wind speed and power performance [26]. With the increased development and installation of wind turbines as wind farms, more work has been investigated in the wake velocity deficits generated by the upfront wind turbines. It has been reported that in a full-wake condition the wind turbine power loss may reach up to 40% [23]. The main objective of all the aerodynamic models is to first determine the induced velocity field generated in the upwind and downwind of the rotor blade where the flow through the wind turbine is considered to be subdivided into several streamtube. Then, the lift and drag coefficient as a function of the incidence angle needed to determine the normal, tangential forces as a function of the azimuth angle and, finally, the torque and the generated power. It is important here to note that the wind turbine performance is affected by many parameters such as wind speed, tip-speed ratio (TSR), airfoil shape and size, turbine aspect ratio (H/R), the solidity of the rotor, the swept area, the rotational speed, and other parameters such as dynamic stall effects, the presence of spoilers. Wind turbine aerodynamic loads and performance predictions in the vortex methods use lifting lines or surface to represent rotor blade trailing and shed vorticity in the wake then; the induced velocity is then determined at any point using the Biot-Savard law [21, 26]. Two types of vortex models have been used in this approach: the fixed-wake and the free-wake models. These vortex models need a significant amount of computer time to predict the aerodynamic loads and performance of the wind turbine machine more accurately than momentum models.

In the models using Navier–Stokes equations such as the case in a steady incompressible laminar flow using finite volume method based on the widely known "SIMPLER" algorithm [27], such models are well suited for wind farms as it can compute the flow velocity everywhere in the rotational plane of the wind turbine machine as well as in its vicinity. In a comparative study conducted by Perić et al. [28] using Blade Element Method and CFD on two types of wind turbines, the DTU 10 MW RWT (Denmark Technical University 10 MW Reference Wind Turbine) blade and the MEXICO blade (Model Rotor Experiments In Controlled Conditions), the authors predict the aerodynamic performance of with an accuracy of 15% accuracy for the 10 MW RWT blade and 6% accuracy for the MEXICO blade experiment data. The author recognizes the CFD's power to provide higher accuracy compared to the BEM or vortex methods. While the BEM was limited to wind speeds of 10 to 12 m/s, the CFD method with $k - \omega$ turbulence model performed the prediction with wind speeds up to a range of 20 m/s. However, the authors indicated that the main disadvantage is the high computational time required for such analysis [28]. The present study will concentrate on the DMS method due to its simplicity, ability to include secondary effects, and, more importantly, the fast computation and run time compared to vortex or numerical models. This model can easily be applied to both VAWT and HAWT.

3. Double, multiple streamtube model (DMS), CARDAAV code

In the Double, Multiple Streamtube models (DMS) developed initially been by Paraschivoiu [11, 29], the variation in the upwind and downwind induced velocities as a function of the azimuthal angle has been included. In this case, the wind turbine blade is divided into several elements assuming no interaction between the components. This method's main principle is to determine the axial and angular induction factors using an iteration method. Then, the forces and power generated are defined in a similar manner to the VAWT. **Figure 1** shows the double streamtube model principle where the machine is represented by a pair of actuator disks in tandem in the upwind and downwind zone, CARDAAV Code. The induced velocity decreases in the axial direction so that the downwind induced velocity is smaller than the upwind zone. With $V_{\infty i}$ representing the local ambient wind velocity, V_e the equilibrium-induced velocity as shown in **Figure 1**, we can write the induced velocities as a function of the induced factor in the upwind zone named a and downwind zone named a'.

$$V = a\, V_{\infty i} \tag{1}$$

$$V_e = (2a - 1)V_{\infty i} \tag{2}$$

$$V' = a'V_e = a'(2a - 1)V_{\infty i} \tag{3}$$

$$V'' = (2a' - 1)(2a - 1)\, V_{\infty i} \tag{4}$$

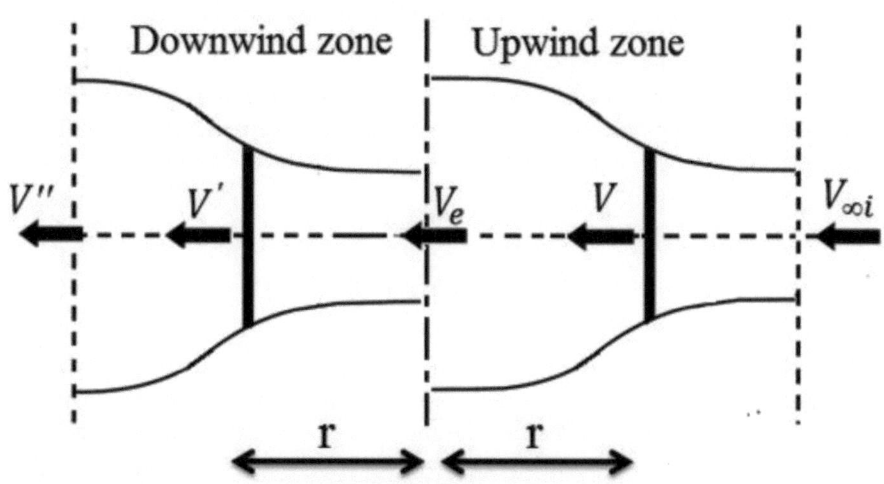

Figure 1.
Principle of the double multiple streamtube model.

The atmospheric wind shear in Eq. (5) is based on the Frost *et al.* model [30] where Z_{EQ} represents the height at the equator, $V_{\infty i}$ is the local ambient wind velocity in the vertical direction, V_∞ is the ambient wind velocity at the equator, and $\alpha_{\infty i}$is the exponent of the power-law, which in our simulation is taken to be 1/7.

$$\frac{V_{\infty i}}{V_\infty} = \left(\frac{Z_{\infty i}}{Z_{EQ}}\right)^{\alpha_{\infty i}} \tag{5}$$

The relative velocity and the local angle of attack in the upstream zone where the azimuth angle varies between $-\pi/2 \le \theta \le \pi/2$, are given by:

$$W^2 = V^2\left[(X - sin\theta)^2 + cos^2\theta\, cos^2\delta\right] \tag{6}$$

$$\alpha = arcsin\left[\frac{cos\theta cos\delta}{\sqrt{(X - sin\theta)^2 + cos^2\theta\, cos^2\delta]}}\right] \tag{7}$$

where $X = \omega r$ is the tip-speed ratio, and ω is the turbine rotational speed. For the downwind zone where the azimuth angle varies between $\pi/2 \le \theta \le 3\pi/2$ similar equations can be derived using W' and α'. The nondimensional normal and tangential forces as a function of the azimuth angle θ are given by

$$F_N(\theta) = \frac{c}{S}\int_{Z_R}^{Z_R+2H}\left(\frac{W}{V_\infty}\right)^2 C_N dZ \tag{8}$$

$$F_T(\theta) = \frac{c}{S}\int_{Z_R}^{Z_R+2H}\left(\frac{W}{V_\infty}\right)^2 \frac{C_T}{cos\delta} dZ \tag{9}$$

Using the same modeling above for the upwind region, the downwind area can also be calculated with the azimuth angle between $\pi/2 < \theta < 3\pi/2$. The normal and tangential force coefficients of the blade section are given by

$$C_N = C_L cos\alpha + C_D sin\alpha \tag{10}$$

$$C_T = C_L sin\alpha - C_D cos\alpha \tag{11}$$

where the lift and drag coefficients C_L and C_D are computed by interpolating the available test data using both the local Reynolds number $\left(R_e = \frac{Wc}{\mu_\infty}\right)$ and the local angle of attack. These coefficients are used up to an angle close to the static stall angle, from which point a dynamic-stall model is considered to estimate the dynamic lift and drag coefficients. The equations for the interference factor $a(\theta)$ are given by [11].

$$K[1 - a(\theta)]cos\theta = a(\theta)f(\theta) \tag{12}$$

$$f(\theta) = \left(\frac{W}{V}\right)^2\left[C_N cos\theta + C_T\left(\frac{sin\theta}{cos\delta}\right)\right] \tag{13}$$

where $K = \frac{8\pi r}{Nc}$ and N is the number of blades. The upwind half-cycle of the rotor is divided into several angular tubes $\Delta\theta$, assuming a constant induced velocity for each of these tubes, **Figure 2**. The interference factor $a(\theta)$ can be written as

$$a(\theta) = \frac{KK_0}{KK_0 + \int_{\theta-\frac{\Delta\theta}{2}}^{\theta+\frac{\Delta\theta}{2}} f(\theta)d\theta} \tag{14}$$

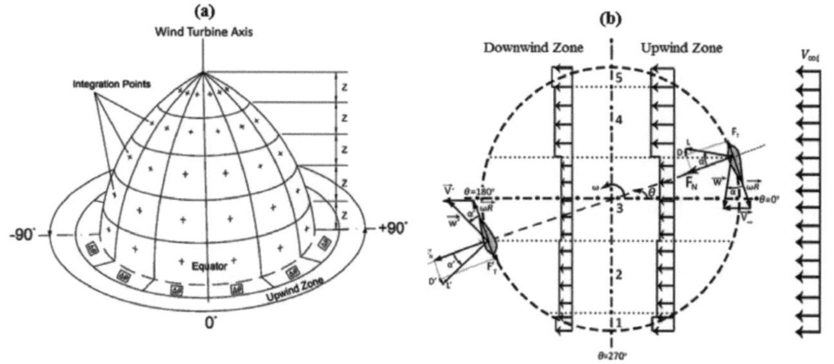

Figure 2.
VAWT model: (a) integration points, (b) DMS model.

where $K_0 = \sin\left(\theta + \frac{\Delta\theta}{2}\right) - \sin(\theta - \Delta\theta/2)$ and $a(\theta)$ is computed numerically. A similar technique is then used for the downwind zone to determine the interference factor $a'(\theta)$. The torque produced by the blade element as a function of the azimuth angle is calculated based on the lift and drag contributions C_{QL} and C_{QD}

$$C_{Q_L} = \frac{Nc}{2\pi SR}\int_{-\pi/2}^{3\pi/2}\int_{Z_R}^{Z_R+2H} C_L sin\alpha\left(\frac{W}{V_\infty}\right)^2 \times \left(\frac{r}{cos\delta}\right) dZ d\theta \tag{15}$$

$$C_{Q_D} = \frac{Nc}{2\pi SR}\int_{-\pi/2}^{3\pi/2}\int_{Z_R}^{Z_R+2H} C_D cos\alpha\left(\frac{W}{V_\infty}\right)^2 \times \left(\frac{r}{cos\delta}\right) dZ d\theta \tag{16}$$

Finally, the power generated due to the lift and the losses due to the drag is given by

$$P_L = \frac{1}{2}\rho_\infty V_\infty^3 S(C_Q)_L \frac{\omega R}{V_\infty} \text{ and } P_D = \frac{1}{2}\rho_\infty V_\infty^3 S(C_Q)_D \frac{\omega R}{V_\infty} \tag{17}$$

The impact of atmospheric wind turbulence on a wind turbine's aerodynamic loads can be included in the DMS model using a stochastic atmospheric wind. The induced velocities necessary for predicting the forces were computed for each Streamtube in both the upwind and downwind zones of the rotor, including the longitudinal and lateral fluctuation velocities resulting from the fluctuating atmospheric wind. A wind time series in the rotor's upwind zone is created to produce the turbulent wind velocity. Using a time delay in the time series based on a linear variation, the turbulent wind speed in the downwind zone is generated [11]. The total velocity of the fluctuating atmospheric wind is a superposition of a mean part and a stochastic fluctuating part. The values of the fluctuations velocities are performed by using the Fast Fourier Transform. The time series are assumed to propagate downstream of the rotor with the speed of the airflow disturbed by the presence of the wind turbine. The simulation uses Double-Multiple Streamtube model (DMS) with up to 1512 stream tubes for the whole Darrieus rotor (21 vertical stream tubes and 72 lateral stream tubes at every five degrees [11]. In the presence of atmospheric turbulence, the relative velocity and the local angle of attack are given by [18]:

$$W^2 = [(V + u_f)\sin\theta + v_f\cos\theta - \omega r]^2 + [v_f\sin\theta - (V + u_f)\cos\theta]^2 \cos^2\delta \tag{18}$$

$$\alpha = \arcsin\left[\frac{v_f\sin\theta - (V + u_f)\cos\theta}{W}\right]\cos\delta \tag{19}$$

where V is the upwind velocity and u_f and v_f the fluctuation velocity components. The overall model developed "CARDAAS" performs the computation of steady-state, CARDAAV as well as the stochastic response of aerodynamic loads when turbulence is included [11, 18, 22].

4. QBlade aerodynamic code

Based on the Blade Element Momentum (BEM) method for the performance prediction of horizontal axis wind turbines (HAWT) and the Double Multiple Streamtube (DMS) method for the performance prediction of vertical axis wind turbine (VAWT), QBlade is an open-source used for wind turbine simulation and distributed under the General Public License with (GPL) free access [31–33]. The code includes extensive post-processing functionality for the rotor and wind turbine simulations, including different rotor blades and variables. The main modules of the QBlade computer code are given in **Table 1** [32]; the modules include the airfoil design, the viscous-inviscid coupled panel method XFOIL analysis, the airfoil polar extrapolation above the stall point, the HAWT and VAWT rotor blade design,

Module		Object
	Airfoil Design	Airfoil
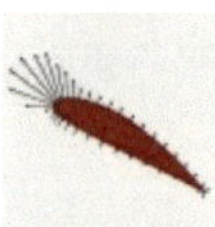	XFOIL Direct Analysis	Polar
360°	Polar Extrapolation to 360°	360° Polar
	HAWT Rotorblade Design	Blade
	Turbine BEM Simulation	Turbine
Q FEM	QFEM - Structural Blade Design and Analysis	Structure
NREL FAST	FAST Simulation	FAST
	Turbulent Wind Field Generator	Wind Field

Table 1.
QBlade objects structure and modules.

the BEM simulation, the structural blade design and analysis, the simulation, and the turbulent wind field generator.

The theoretical formulation in QBlade is based on BEM and DMS. The rotor blade is discretized into a finite number of blade elements with defined cross-sections according to the radial position, profile, chord, twist, and length. Using the momentum theory, each blade section's relative wind speed is computed, which is then used to calculate the angle of attack and the Determination of the airfoil lift and drag coefficients. Once these parameters are known, the aerodynamic normal and tangential forces are computed, then the thrust and torque of an element are determined. Similar to CARDAAV code, in QBlade, the iteration variables of the BEM method are the two induction factors, axial a, and radial a' as shown in **Figure 3** where α is the angle of attack, $\varnothing$ is the inflow angle, θ is the pitch angle, and β is the twist angle. The wind velocity at the rotor blade is given by $V(1-a)$ in the horizontal direction, and the angular velocity is given by $\Omega r(1+a')$.

Based on **Figure 3**, the inflow angle $\varnothing$ and the relative velocity V_R are given by the equations

$$\varnothing = tan^{-1}\left[V_\infty \frac{1-a}{\Omega r(1+a')}\right] \tag{20}$$

$$V_R = V_\infty \frac{1-a}{sin\,\varnothing} \tag{21}$$

Using the momentum and BEM theory and the solidity $\sigma_r = \frac{Bc}{2\pi r}$ with B the number of blades and c the chord length, the axial and radial induction factors a and a' are computed from

$$a = \frac{1}{\frac{4\sin^2\varnothing}{\sigma_r C_N}+1} \text{ and } a' = \frac{1}{\frac{4\,sin\,\varnothing\,cos\,\varnothing}{\sigma_r C_T}-1} \tag{22}$$

where the normal and tangential force coefficients of the blade section are given by

$$C_N = C_L\,cos\,\varnothing + C_D\,sin\,\varnothing \tag{23}$$

$$C_T = C_L\,sin\,\varnothing - C_D\,cos\,\varnothing \tag{24}$$

The iteration technique used in the above equations is first to initialize the axial and radial induction factor a and a', then computer the inflow angle from Eq. (20),

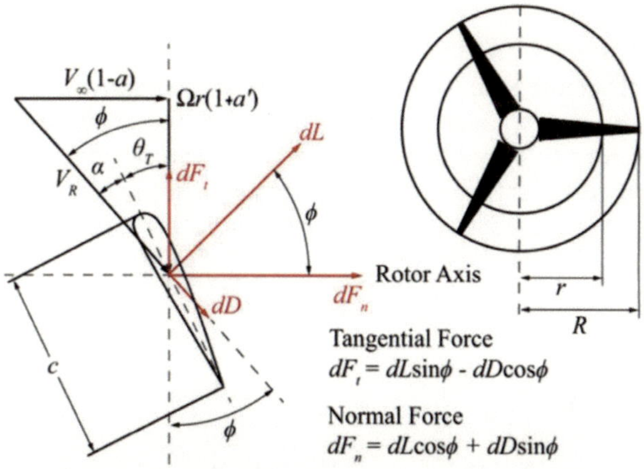

Figure 3.
Angles, velocity components, and forces acting on the HAWT rotor blade section.

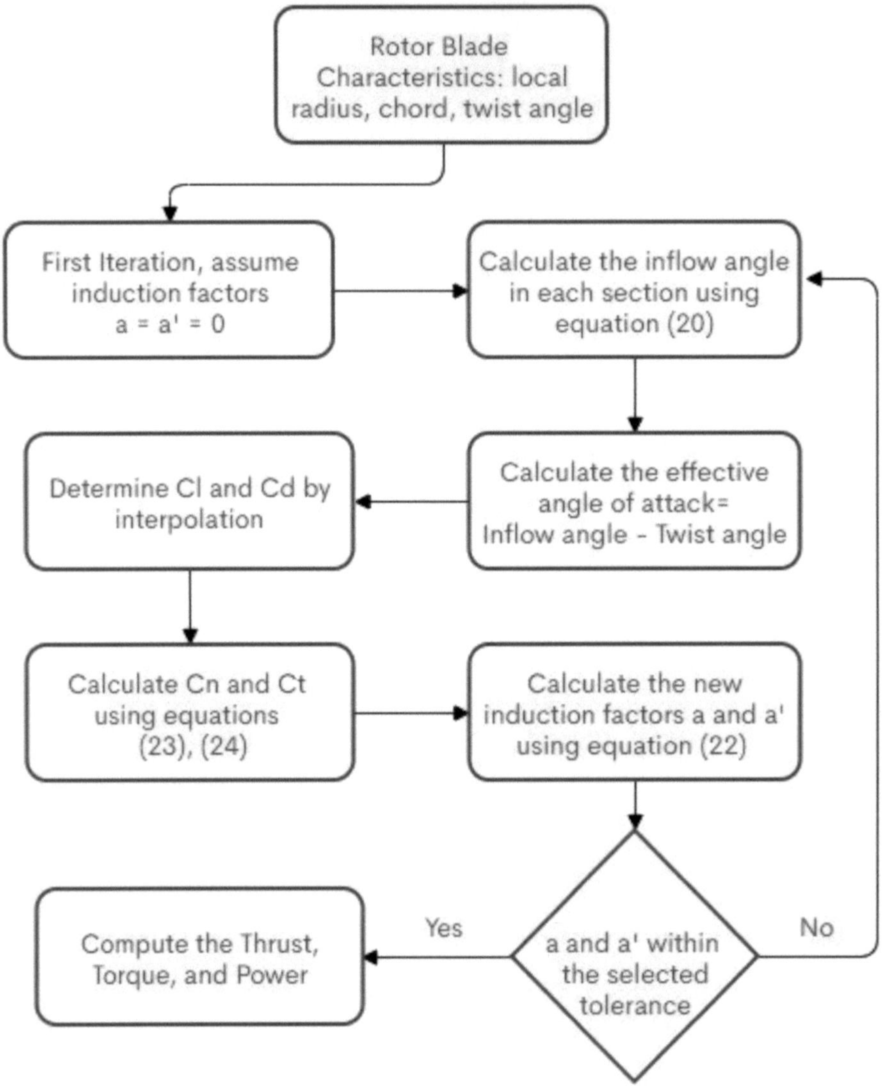

Figure 4.
Flowchart of the iterative algorithm used in QBlade.

the local angle of attack by subtracting the twist angle from inflow angle, use aerodynamic coefficients from tabulated airfoil data, compute a and a' from Eq. (22), compare the new axial and radial induction factors with the previous one, if not satisfied, return to step 2 by recalculating the inflow angle again in Eq. (20) and repeat the process if satisfied compute aerodynamic loads and performance of the wind turbine. For the vortices that form at the rotor's tip, resulting in added drag, the Prandtl tip loss factor is introduced [29, 33]. **Figure 4** shows a flowchart of the algorithm used in QBlade.

5. Secondary effects

The computer code CARDAAV can evaluate several rotor shapes with a straight or curved blade and use a defined shape such as a parabola, catenary, or modified troposkien. The code also can include the so-called "secondary" Effects," such as the rotating tower, strut, and the spoiler. Another involved and unsteady phenomenon due to airfoil undergoing large and rapid variations of the angle of attack with time is the dynamic stall. During the rotor's rotation, the drag and lift coefficients present a different hysteresis than the case of static behavior. A dynamic delay of the stall to angles is substantially beyond the static stall angle, including massive recirculating regions moving downstream over the airfoil surface. In the case of VAWT, when the operational speed approaches its maximum, all the rotor blade sections go beyond the critical static stall angle of attack, the angle of attack changes substantially, and the entire blade operates under dynamic stall conditions, which will increase the unsteady blade loads and the wind machine structural fatigue. Different dynamic stall models with some modifications derived from static and dynamic airfoil tests have been incorporated into the CARDAAV computer code, such as the Gormont model and the indicial model [11, 34, 35]. Comparisons between aerodynamic performance predictions using dynamic-stall models show that the models provide a better prediction of the dynamic-stall regime characterized by a plateau oscillating near the experimental data of the rotor power function of wind speed [11].

6. Result and discussion

This section presents a selection of results obtained by performing aerodynamic loads and performance prediction using CARDAAV, different variants, and QBlade computer codes, including dynamic stall. Results were achieved on Sandia 17-m and 34 wind turbine machine [36] and compared to available experimental data. CARDAAV is the original code using an ambient constant ambient atmospheric wind speed, while CARDAAS code incorporated a stochastic wind to account for the atmospheric turbulence. Both CARDAAV and CARDAAS are based on DMS methods.

The viscous code 3DVF uses numerical methods to simulate the flow field of VAWTs in cylindrical coordinates based on the solution of steady, incompressible, and laminar Navier–Stokes equations. Different dynamic stall models have been incorporated into the aerodynamic codes, namely the original Gormont model, the

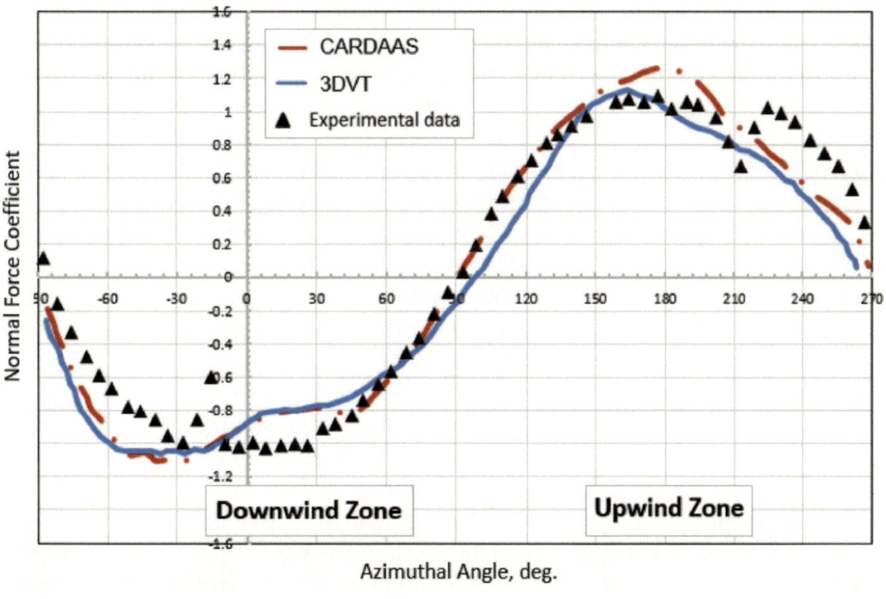

Figure 5.
Normal force coefficient as a function of azimuthal angle for 17-m wind turbine machine at TSR = 2.86.

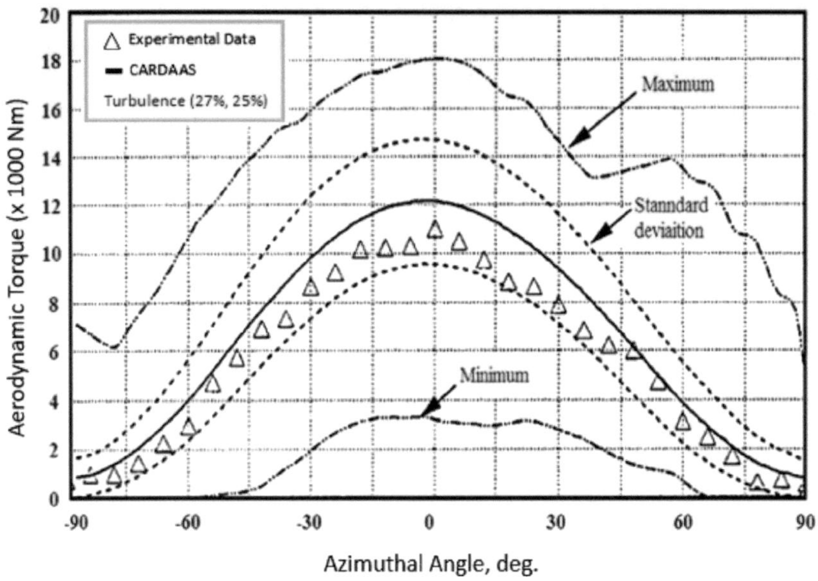

Figure 6.
Comparison of aerodynamic torque with experimental data at TSR = 4.60 and 50.6 rpm.

indicial model, and the MIT model [11, 34, 37]. **Figure 5** compares the normal force coefficient's distribution as a function of azimuthal angle for 17-m wind turbine machine at TSR = 2.86 using CARDAAS and 3DVF codes and experimental data. Using CARDAAS with low turbulence intensity induces a low variation in the ensemble-averaged values compared to the high turbulence intensity; obviously, for 0 turbulence intensity, both CRDAAV and CARDAAS give the same results. Comparing the two codes, one can conclude that CARDAAS and 3DVF predict quite well the distribution of the normal force coefficient. However, 3DVF is more time-consuming. **Figure 6** compares the normal force coefficient's distribution as a function of azimuthal angle for a 17-m wind turbine machine at TSR = 4.60 for a rotor rotation of 50.6 rpm. The distribution of the torque compares quite well with the experimental data. At the same time, it also gives the possibility to estimate the maximum and minimum torque encountered during the wind turbine rotation as a function of the azimuthal angle. **Figure** 7 compares the wind turbine's theoretical performance with the experimental data for the Sandia 17-m turbine. The original CARDAA code does not include the variable interference factors used in CARDAAV code. As shown in this figure, CARDAA over predicts the power coefficient peak while CARDAAV shows a relatively good agreement with experimental data due to dynamic and secondary effects. Different dynamic stall models can be incorporated into the available aerodynamic codes. In **Figure 8**, the Gormont dynamic stall model and the improved Gormont model, along with the MIT dynamic stall models, have been used to predict the power generated by the wind turbine at different wind speeds.

All in all, the models forecast well the power generated by the wind turbine, but the early version of the Gormont over predict that power. In fact, the improved Gormont model by Paraschivoiu [11] takes into account the fact that high-level turbulence delays the onset of the dynamic stall. Based on that, the improved Gormont dynamic stall model is used at low turbulence zone only, namely between an azimuthal angle of 135 degrees and 15 degrees; the rest of the azimuthal angle will ignore the dynamic stall. In the MIT model case, the dynamic-stall regime is characterized by some variation compared to the improved Gormont model. In **Figure 9**, we present the case of the QBlade code used for HAWT and VAWT.

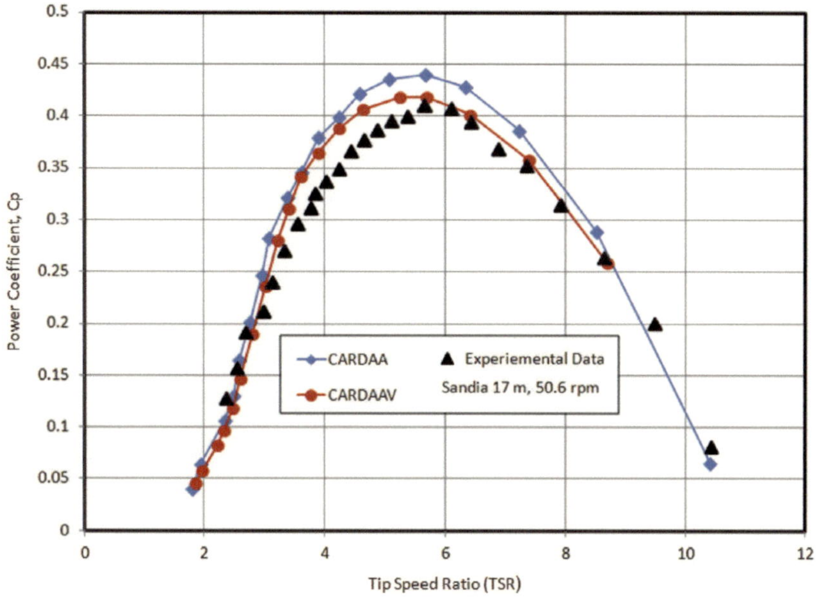

Figure 7.
Performance comparison between theoretical and experimental data for the Sandia 17-m turbine.

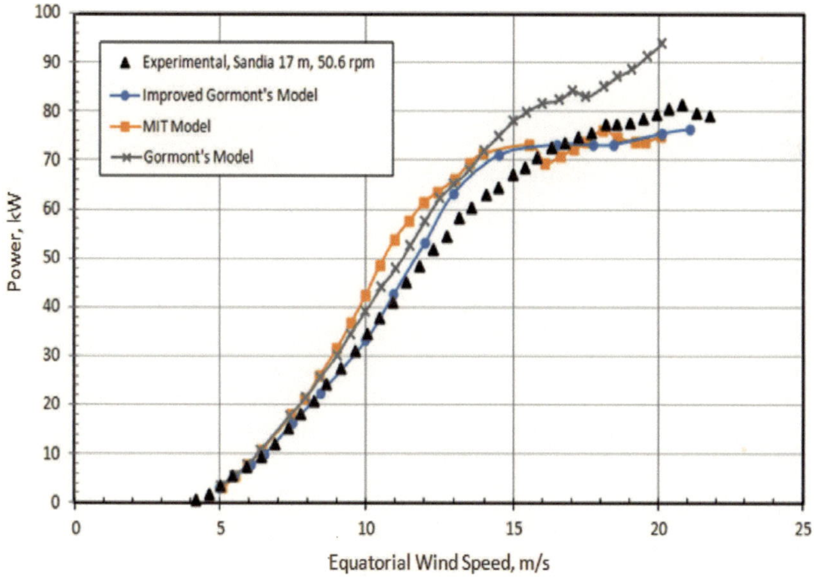

Figure 8.
Comparison of rotor power with different dynamic stall models at 50.6 rpm.

QBlade also uses the Blade Element Momentum (BEM) method and the DMS method to simulate HAWT and VAWT performance. Details on how to use QBlade can be found [31–33]. As an example, we generated the SG6043 airfoil pressure distribution generated using the XFOIL QBlade module. QBlade code can simulate the wake in the "Turbine Visualization" module, as shown in **Figure 10**. In **Figure 11**, we compare QBlade with CARDAAV code, including two dynamic stall models for the Sandia 34-m turbine, namely the indicial and improved Gormont models. The indicial model is used to simulate the effect of a dynamic stall at low

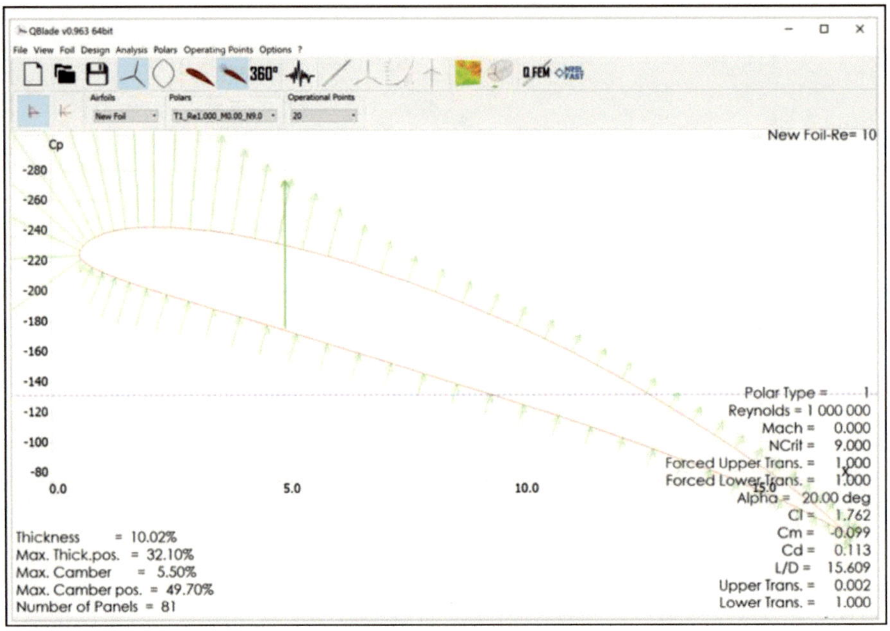

Figure 9.
Example of the SG6043 airfoil pressure distribution generated by QBlade.

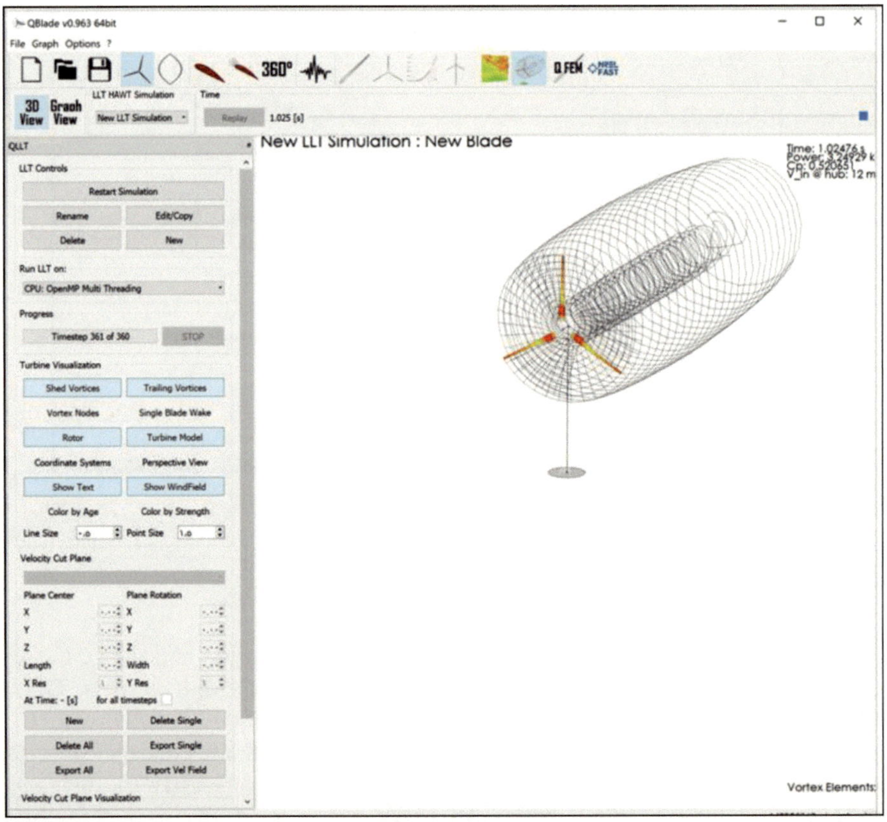

Figure 10.
Example of the wake simulation generated by QBlade.

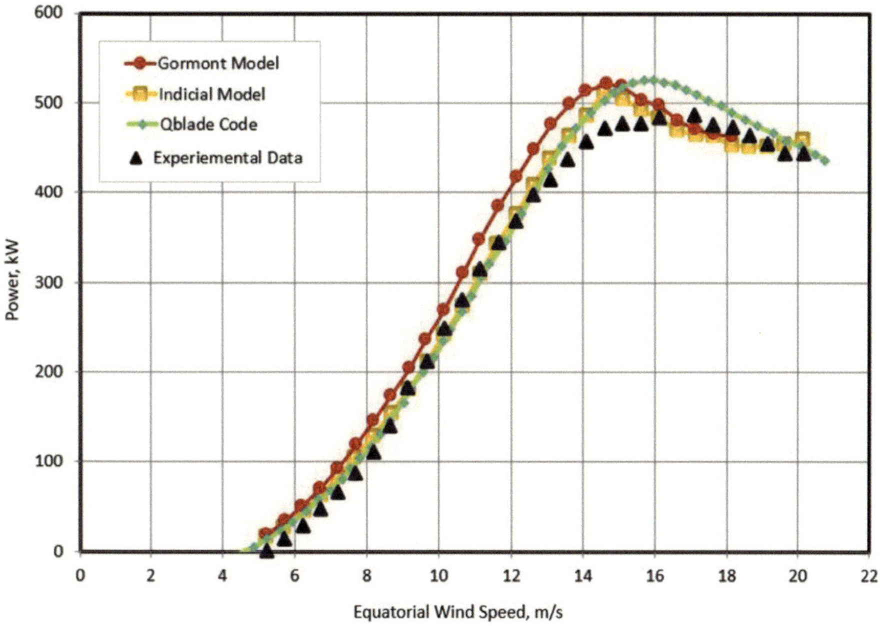

Figure 11.
Power distribution for the Sandia 34-m turbine using indicial and improved Gormont models at 34 rpm.

tip-speed ratios; it is basically centered on the fact the dynamic stall is considered to be a superposition of several different effects that can be independently explored by using indicial functions [34]. As shown by **Figure 11**, all models predict quite well the power generated by the wind turbine but QBLADE slightly over predicts the power compared to CARDAAV. More validation of the QBLADE code in the case of VAWT can be found in [38].

7. Conclusion

This book chapter emphasizes progress in the development of the aerodynamic codes for the prediction of HAWT and VAWT such as CARDAAV and Qbalde, with a focus on the BEMN and DMS methods for predicting aerodynamic and performance of wind turbines, including normal and tangential forces, torque, and the power generated by the wind turbine. The dynamic stall has been incorporated in this study to show the effects of this phenomenon on wind turbines' performance. The codes have been successfully applied compared to available experimental data for the 17 m and 34 m wind turbine machines. The strength of the available codes lies in their simplicity, fast prediction, and ability to include secondary effects. More work is now needed to simulate wind turbine wake and predict the performance of wind turbine farms, which requires the understanding of the physics of the atmospheric wind flow, the structure and dynamics of large-scale machines, and the enhancement of energy capture, control, stability, optimization, and reliability.

Acknowledgements

The authors gratefully acknowledge the support of the College of Engineering at Effat University, Jeddah, the Kingdom of Saudi Arabia, and the J.- Armand Bombardier Chair at Polytechnique of Montreal, Canada.

Author details

Tayeb Brahimi[1]* and Ion Paraschivoiu[2]

1 Energy Research Lab, College of Engineering, Natural Science, Mathematics and Technology Unit Effat University, Jeddah, Saudi Arabia

2 Department of Mechanical Engineering, Polytechnique Montreal, Montreal, PQ, Canada

*Address all correspondence to: tbrahimi@effatuniversity.edu.sa

References

[1] AWEA. (2020, April 16). *American Wind Energy Association. Q1 2020 Installed Wind Power Capacity (MW)*. WINDExchange. https://windexchange.energy.gov/maps-data/321

[2] WWEA. (2020). World wind capacity at 650,8 GW, Corona crisis will slow down markets in 2020, renewables to be core of economic stimulus programmes. In *World Wind Energy Association*. https://wwindea.org/blog/category/statistics/

[3] GWEC. (2020). Global Wind Energy Report 2019. In *Global Wind Energy Council*. https://gwec.net/global-wind-report-2019/

[4] WindEurope. (2020). Wind energy in Europe in 2019 Trends and statistics. In *WindEurope*. Wind Europe. https://windeurope.org/about-wind/statistics/

[5] EIA. (2017). *Annual Energy Outlook 2017 with projections to 2050* (Report, U.S. Energy Information Administration #AEO2017; Annual Energy Outlook). U.S. Energy Information Administration. https://www.eia.gov/outlooks/aeo/

[6] IEA. (2020). *Renewable energy market update—IEA*. https://www.iea.org/reports/renewable-energy-market-update/covid-19-impact-on-renewable-energy-growth

[7] Sönnichsen, N. (2020). Covid-19 revised global wind energy additions forecast 2020. In *Statista*. https://www.statista.com/statistics/1107540/covid-19-global-wind-power-additions-forecast/

[8] Saurabh, A. (2020, May 23). *Nearly 85% Of Power Capacity Added in India in Q1 2020. Solar & Wind*. https://cleantechnica.com/2020/05/23/nearly-85-of-power-capacity-added-in-india-in-q1-2020-was-from-solar-wind/

[9] Veers, P., Dykes, K., Lantz, E., Barth, S., Bottasso, C. L., Carlson, O., Clifton, A., Green, J., Green, P., Holttinen, H., & al, et. (2019). Grand challenges in the science of wind energy. Science, *366* (6464). https://doi.org/10.1126/science.aau2027

[10] Kuik, G. A. M. V., Peinke, J., Nijssen, R., Lekou, D., Mann, J., Sørensen, J. N., Ferreira, C., Van Wingerden, J. W., Schlipf, D., Gebraad, P., & al, et. (2016). Long-term research challenges in wind energy – a research agenda by the European Academy of Wind Energy. *Wind Energy Science*, *1*(1), 1–39. https://doi.org/10.5194/wes-1-1-2016

[11] Paraschivoiu, Ion. (2009). *Wind turbine design: With emphasis on Darrieus concept*. Presses internationales Polytechniques.

[12] Sørensen, J. N. (2016). *General momentum theory for horizontal axis wind turbines*. Springer.

[13] *Deployment, investment, technology, grid integration and socio-economic aspects*. (n.d.). Retrieved January 16, 2021, from https://www.polity.org.za/article/deployment-investment-technology-grid-integration-and-socio-economic-aspects-2019-10-22

[14] Amano, R. S. (2017). Review of Wind Turbine Research in 21st Century. Journal of Energy Resources Technology, *139*(050801). https://doi.org/10.1115/1.4037757

[15] *Wind turbine rotor diameter size 2021*. (n.d.). Statista. Retrieved January 16, 2021, from https://www.statista.com/statistics/860565/global-rotor-diameter-size-of-wind-turbines/

[16] Reed, S. (2021, January 1). A Monster Wind Turbine Is Upending an Industry. *The New York Times*. https://

www.nytimes.com/2021/01/01/business/GE-wind-turbine.html

[17] Paraschivoiu, Ion, & Allet, A. (1988). Aerodynamic analysis of the Darrieus wind turbines including dynamic-stall effects. Journal of Propulsion and Power, *4*(5), 472–477. https://doi.org/10.2514/3.23090

[18] Brahimi, T., & Paraschivoiu, I. (1995). Darrieus Rotor Aerodynamics in Turbulent Wind. *1995*, *117*(2), 128–136. https://doi.org/10.1115/1.2870839

[19] Doubrawa, P., Barthelmier, R., Wang, H., & Chrchfield, M. (2016). A stochastic wind turbine wake model based on new metrics for wake characterization. WInd Energy, *20*(3). https://doi.org/10.1002/we.2015

[20] Branlard, E. (2017). The Blade Element Momentum (BEM) Method. In E. Branlard (Ed.), *Wind Turbine Aerodynamics and Vorticity-Based Methods: Fundamentals and Recent Applications* (pp. 181–211). Springer International Publishing. https://doi.org/10.1007/978-3-319-55164-7_10

[21] Hansen, M. O. L., Sørensen, J. N., Voutsinas, S., Sørensen, N., & Madsen, H. Aa. (2006). State of the art in wind turbine aerodynamics and aeroelasticity. Progress in Aerospace Sciences, *42*(4), 285–330. https://doi.org/10.1016/j.paerosci.2006.10.002

[22] Paraschivoiu, I., Saeed, F., & Desobry, V. (2002). Prediction Capabilities in Vertical-Axis Wind Turbine Aerodynamics. *, Berlin, Germany, 2–6 July 2002.*

[23] Sanderse, B., Pijl, S. P. van der, & Koren, B. (2011). Review of computational fluid dynamics for wind turbine wake aerodynamics. Wind Energy, *14*(7), 799–819. https://doi.org/10.1002/we.458

[24] Sørensen, Jens Nørkær. (2011). Aerodynamic Aspects of Wind Energy Conversion. Annual Review of Fluid Mechanics, *43*(1), 427–448. https://doi.org/10.1146/annurev-fluid-122109-160801

[25] Tristan, D., Turaj, A., & Joaquim, M. (2016). Aerodynamic shape optimization of wind turbine blades using a Reynolds-averaged Navier–Stokes model and an adjoint method. Wind Energy, *20*(5). https://doi.org/10.1002/we.2070

[26] Brahimi, T. (2019). Using Artificial Intelligence to Predict Wind Speed for Energy Application in Saudi Arabia. Energies, *12*(24), 4669. https://doi.org/10.3390/en12244669

[27] Allet, A., Brahimi, M. T., & Paraschivoiu, I. (1997). On the aerodynamic modelling of a VAWT. Wind Engineering, *21*(6), 351–365.

[28] Perić, B. M., Simonović, A. M., & Vorkapić, M. D. (2006). Comparative analysis of numerical computational techniques for Determination of the wind turbine aerodynamic performances. Thermal Science JOurnal, *2*. https://doi.org/10.2298/TSCI200216175P

[29] Hanson, M. O. L. (2008). *Aerodynamic of wind Turbines* (2nd ed.). Earthscan Publications Ltd.

[30] *NASA Technical Reports Server (NTRS)*. (n.d.). Retrieved January 18, 2021, from https://ntrs.nasa.gov/citations/19790006507

[31] Marten, D., Wendler, J., Pechlivanoglou, G., Nayeri, C. N., & Paschereit, C. O. (2013). QBlade: An open source tool for design and simulation of horizontal and vertical axis wind turbines. International Journal of Emerging Technology and Advanced Engineering, *3*(3), 264–269.

[32] *QBlade*. (n.d.). Retrieved January 18, 2021, from http://www.q-blade.de/

[33] HFI TU Berlin. (2013). QBlade Wind Turbine Design and Simulation. In *QBlade*. http://www.q-blade.org/

[34] Allet, A., & Paraschivoiu, I. (1995). *Viscous Flow and Dynamic Stall Effects on Vertical-Axis Wind Turbines*. International Journal of Rotating Machinery; Hindawi. https://doi.org/10.1155/S1023621X95000157

[35] Masson, C., Leclerc, C., & Paraschivoiu, I. (1998). *Appropriate Dynamic-Stall Models for Performance Predictions of VAWTs with NLF Blades*. International Journal of Rotating Machinery; Hindawi. https://doi.org/10.1155/S1023621X98000116

[36] Worstell, M. H. (1979). *Aerodynamic performance of the 17-metre-diameter Darrieus wind turbine* (SAND-78-1737). Sandia Labs., Albuquerque, NM (USA). https://www.osti.gov/biblio/6531371-aerodynamic-performance-metre-diameter-darrieus-wind-turbine

[37] Khan, M. A. (2018). *Dynamic Stall Modelling for Wind Turbines*. https://doi.org/10.13140/RG.2.2.26000.35848

[38] Marten, David, Lennie, M., Pechlivanoglou, G., Paschereit, C., Dy, N., Paraschivoiu, I., & Saeed, F. (2017). Validation and comparison of a newly developed aeroelastic design code for VAWT.

Chapter 8

Assessment and Analysis of Offshore Wind Energy Potential

Radian Belu

Abstract

Wind energy usage is increasing at fast rates due to significant technical advances, energy supply security and environmental concerns. Research is focusing among others areas on the development of reliable and accurate wind energy assessment methods. Offshore wind energy resources are usually larger than at geographically nearby onshore sites, which may offset in part higher installation, operation, and maintenance costs. Successful offshore wind energy development relies on accurate analysis and assessment of wind energy resource potential. Offshore wind assessment challenges are related to the wind turbine size, offshore installation challenges, lack of adequate and long-term wind and meteorological measurements, etc. Wind, a highly intermittent phenomenon has large spatiotemporal variability, being subject to sub-hourly, hourly, diurnal, seasonal, yearly, and climate variations in addition to their dependence on the geography and environment. Wind regime characteristics are critical to all aspect of a wind energy project, e.g. potential site identification, economic viability, equipment design, operation, management, or wind farm impacts on the electric grid. For a reliable wind energy assessment, measurements at rotor heights are required at least for one year. If such measurements are not available needs to be substituted by alternative approaches, e.g. measure-correlate-predict or numerical methods. Chapter objectives are to provide the reader with comprehensive reviews of the wind energy assessment and analysis methods.

Keywords: wind regime, offshore resource assessment, wind characteristics, wind profiles, turbulence intensity, wind statistics, measure-correlate-predict techniques

1. Introduction

Wind power is viewed as one of the most techno-economically viable renewable energy sources for electricity generation. In resource-ideal locations, the wind generated electricity costs are competitive with conventional power generation. Accurate estimates of the energy production together with good estimates of the uncertainties associated with any project are required to secure funds and hedge wind project risks. Wind energy resource assessment enters into the several project development phases: (1) suitable wind energy site prospecting, (2) site mapping and wind farm design, (3) wind turbine micro-siting, (4) risk assessment and performance analysis, (5) permitting and tower certification, and (6) wind farm operation and management [1–21]. Better knowledge, the smaller are the safety margins, therefore, the higher investment potential returns. An accurate prediction of the potential wind farm performance is therefore vital for the project success.

Electricity generation from wind can be economically achieved only where a significant wind resource exists. To maximize the wind form power output, the resource assessment at any prospective site is critical. Wind is highly variable, geographically and temporally, its variability persisting over a very wide range of spatiotemporal scales. Due to the cubic relationship between wind speed and energy output, sites with small differences in wind speeds can have substantial differences in available wind energy [2–10]. A wind power assessment is accurate, only if the wind speeds and directions are measured at the wind turbine hub height and for a significant time-period. Knowledge of the local wind regime is vital to the industry, yet commercially viable products that meet the wind industry needs are often questionable. Three main phase involved with project planning and operation, requiring accurate wind characterization are: (1) prospecting and siting, by using historical data, retrospective forecasts, and statistical methods to identify potential wind energy sites; (2) specific micro-siting assessment to determine the optimum wind energy project layout; and (3) operation using the wind prediction to determine available power output for specific time horizons. However, the most critical factor is the wind energy resource identification and characterization. Wind energy resources depend on the wind regime, varying in time and space due to large- and small-scale atmospheric circulations, surface energy fluxes, and geography. Ultimately, wind energy production is governed by factors such as: large-scale generation potential, grid supplied power predictability, and the expected investment returns. The various wind energy uncertainties impact the reliable determination of these viability factors.

Offshore wind power relates to the installation of wind turbines in large water bodies. On average, winds blow faster and more uniformly over the sea than on land, and a faster and steadier wind means less wear on the turbine components and more electricity generated per turbine [12–20]. Due to cubic relationship of the potential energy produced from wind and the wind speed, any wind speed marginal increase results in larger generated electricity. Notice that the offshore wind energy is also the most developed form of marine renewable energy in terms of technology, policy frameworks, and installed capacity. The offshore wind industry needs detailed wind regime information for proper structural design, e.g. wind shear and veer, across the rotor plane as well as between the water surface and hub height, turbulence, ideally at hub height, wind speed profile, wind velocity probability distributions, and extreme weather conditions. In addition to detailed wind information, other environmental data are also important during the design and operation, e.g. air and water temperature and gradients, tidal, storm surge, extreme waves, marine currents, atmospheric humidity, pressure, density, icing characteristics, hail and lightning frequency and severity, and seismic conditions [2–4, 9–22].

The most important activity in a site selection is to determine the wind energy resource potential, consisting in the estimated local wind velocity probability density function. Other important aspects in this context are estimating the turbulence levels and the resulting wind turbine loads at the concerned site, promoting better decision making, in selecting the most suitable site wind turbines and the project life cycle cost prediction. Higher wind loads result in higher maintenance and operation costs [1–27]. Other site selection criteria are: (1) local geography, (2) electric grid proximity, (3) permitting and land acquisition, and (4) site accessibility for transportation and maintenance. A planning strategy accounting for key engineering design factors and addressing the uncertainties in the wind energy project offers to the wind energy industry a powerful impetus. However, the wind energy resource itself is highly uncertain, while the wind conditions (wind speed and direction), turbulence intensity and air density are showing large temporal variations, varying significantly from season to season or from year to year.

Inaccuracies in wind velocity distributions can introduce significant errors in the estimated wind resource potential or in the wind farm performance prediction. Key measures of wind power plant performance include annual energy production, cost of energy, and payback period. Both parametric and nonparametric uncertainty models are formulated, which can be leveraged in conjunction with a wide variety of wind velocity distribution models [2–14].

Wind energy measurement campaigns have traditionally been conducted with mast-based instrumentation consisting of cup anemometers and wind vanes. Almost all field campaigns are also using temperature, pressure, and humidity sensors, besides the wind velocity sensors [2–16]. However, many wind energy developers and manufacturers stick to the standard mast configuration for long-term measurement campaigns. Notice that sonic anemometers and wind remote sensing instruments are still far from being standard instruments in wind energy assessment although they are being used more and more for detailed measurement campaigns or performance testing. Given the field campaign limited time span, it is often needed to extrapolate wind time series to periods of at least 5 years to better predict the long-term average energy yield. For this purpose, the measure-correlate-predict (MCP) methods are still the most used methods. Numerical models are especially used in wind resource assessment to spatially extrapolate the wind measurements to obtain wind maps, and vertically to estimate hub-height wind fields. Additionally, numerical weather prediction models are used to construct simulated historical time series that can be used to extend limited wind measurement time series to longer time spans in a similar way to the MCP methods. Usually measurement techniques are available for on-site measurement ranging from point measurements performed at different heights using anemometers or ultrasonic sensors to profiling techniques like sonic detection and ranging (SODAR) or light detection and ranging (LIDAR) systems. However, many measurement campaigns for commercial wind projects or even for wind energy research projects rely on cup anemometry and ultrasonic sensors, the latter being preferred in research projects. However, remote-sensing techniques (SODARs and LIDARs) are increasingly used as complementary techniques, providing high quality wind vertical profiles at higher sampling rates [1–20]. In wind projects the profiling instruments can be conveniently relocated within the project area for specific wind measurements. Remote sensed wind speed measurements are needed to supplement mast measurements, especially in off-shore, campaign to evaluate wind flow models for resource assessments, power curve measurements, or uncertainty evaluation.

The primary objectives of robust, accurate and optimal wind farm planning include: optimal site selection, based on the quality of the wind energy resource, maximization of the annual energy production, energy cost minimization, and reliability wind farm energy production maximization. Major activities in a site selection are accurately determining the wind energy potential at the candidate site, turbulence levels and the resulting wind loads at the wind farm site. Such activities are critical for selecting site optimum wind turbines and to predicting the life cycle cost of the project, higher wind loads usually implying higher costs. Other wind energy site selection criteria include, but not limited local geography, distance to electric grid connections, permitting, and site accessibility.

1.1 Issues and challenges of the offshore wind resource assessment

Developing any offshore wind energy project presents unique and specific challenges, different from the onshore wind energy counterparts. Offshore wind farms are subject to specific atmospheric conditions, the see atmosphere boundary layer dynamics are significantly different from the land ones, and in many regards, not fully understood [2–14]. Moreover, the atmospheric observations over the ocean or see are sparse compared to land, requiring an increasing reliance on numerical or MCP models to assess wind energy potential [2–4, 9–18]. Good wind energy potential is by far the most important factor, in any offshore wind energy project, considering relative merits of the potential sites. Wind energy resource assessment, covering a wide range of spatiotemporal scales, is playing a critical role in determining the wind energy project economic viability, described as a two-phase process: (a) regional wind energy resource assessment, and (b) site specific analyses of the energy resource quality. Successful offshore wind energy projects are relying on accurate estimate of the wind regime. However, since installing and operating offshore measurement equipment and meteorological masts are expensive and difficult to operate; prospective sites must be carefully evaluated through remote sensing, numerical model data or measure-correlate-predict approaches [2–4, 9, 14–18]. Relative paucity of comprehensive offshore wind velocity observations, with proper spatio-temporal resolution makes the offshore wind energy assessment more challenging than for the onshore sites. Notice also that the floating offshore wind technology is constrained by some practical installation depths (up to 1300 m depth), limiting the offshore wind energy availability and areas. The depth limits up to 1300 m are based on economic criteria, safety and installation issues for the electrical undersea power cables.

There are main issues regarding the offshore wind energy resources, e.g. atmospheric stability conditions, wind dynamics, the extrapolation of the wind speed at turbine hub heights due to the lack of adequate observations. For example, the influence of thermal stratification on vertical profiles of wind speed is believed to be larger than over land due to lower mechanically generated turbulence [9, 14]. Other issues affecting offshore wind energy resources are that many of the offshore wind farms are located in the coastal areas, the region extending from the coastline, where the wind velocity and turbulence profiles are not in equilibrium with the underlying sea surface, significantly affecting the wind shear, wind profiles and turbulence. Research works are suggesting that the distance from the coastline over which wind speed vertical profiles are not at equilibrium with the sea surface extends for about 20 km and possibly larger distance from the coastline [9, 12–14]. Several studies have demonstrated that careful area meteorology and climatology considerations, when determining the layout of an offshore wind farm can increase its power production, improving wind farm viability [9, 12–16]. It was shown by using a mesoscale model, incorporating a wind farm parameterization can improve wind energy resource assessment [9, 13–16]. Offshore wind energy assessments should also take into consideration the experience, technology advancements, and trends of the offshore wind industry over the past few decades to establish physical parameters for array power density and wind turbine height that are needed to accurate evaluate the power capacity and energy production [2–4, 9–19]. Notice that the assessment of availability for good offshore wind energy sites is dependent on meteorological see conditions, the service equipment availability, electric grid proximity and the land-based infrastructure.

2. Factors affecting wind power computation

Wind shear, turbulence intensity, and atmospheric stability effects on wind turbine production are not fully understood, and can introduce large uncertainties on wind energy assessment. The estimation of these uncertainties is related to empirical considerations rather than theoretical calculations. Several studies are suggesting that the natural variability of wind energy resources should include air density, surface roughness, associated probability distributions, and error for prediction of long-term wind velocity. Depending on atmospheric conditions, waking by upstream wind turbines and roughness interactions, wind turbines often operate far from the ideal conditions, field-deployed power curves being quite different from certified ones [2–12]. Better predictions of power output or loads require representative wind measurements and power estimates over the rotor-swept area for individual wind turbines. Depending on the flow properties and motion scales, the flow can become turbulent [2–4]. There are three approaches that can reduce the wind energy intermittency: spatial distribution of wind power facilities, accurate forecasting, and energy storage systems. Although wind generation is subject to large wind variations, if the facilities are spatially distributed, an overall output at any time is more uniform and reliable. The wind speed increases with height, higher elevation sites offering greater wind energy resources. However, the air density decrease with height reduces wind turbine power output. However, it is advantageous to locate turbines at higher elevations to take advantage of higher wind speeds. Power curves for various air density values must be accounted for better power output estimate accuracy. Air density is usually calculated from temperature and pressure measurements [2–4]. Depending on the wind turbine, either the wind velocity is normalized for power calculations. The wind speed is normalized with the reference air density ρ_0, as:

$$v_{norm} = \overline{v}\left(\frac{\overline{\rho}}{\rho_0}\right)^{1/3} \tag{1}$$

The usual hub-heights of 80 m or higher of the modern, the rotors may encounter large vertical gradients of wind speed and boundary layer turbulence. Wind turbine rotors are susceptible to turbulence fatigue damages. Understanding of the turbulence impact on the blades can help in better designing the operational and maintenance schedules for wind farms. Consequently, the full understanding can lead to advanced and improved control and management schemes and methods. Quantification of the turbulence effects on wind turbine is usually done by computing an equivalent fatigue load parameter, as a function of wind fluctuation amplitudes within the averaging period, blade material properties, and the size of measurement samples. It is found that the highest blade root flap bending moment equivalent fatigue load does not correspond to the greatest wind speeds, but to the class of wind speeds that has the highest amplitude of the fluctuations [2–4, 18–27]. Turbulent fluctuations are found to be the main source of the blade fatigue. The turbulence intensity (TI), a measure of the overall level of turbulence, is defined as:

$$TI = \frac{\sigma_v}{v} \tag{2}$$

where σ_v is the wind speed standard deviation (m/s) at the nacelle height over a specified averaging period (usually 10 min). In [21] was found that, from the power curves for different turbulence intensity classes and for low to moderate wind speeds (4 to 12 m/s), the high TIs are yielding the higher turbine power output. TI index is affected by the atmospheric stability, so the theoretical wind turbine power curves [2–4, 13–21]. A correction factor, common used, for the effect of turbulence intensity is given by:

$$v_{corr} = v_{norm}\left(1 + 3(TI)^2\right)^{1/3} \tag{3}$$

Vertical wind shear effects are important as the wind turbines become larger and larger. It is therefore quite questionable whether the hub height wind speed is representative. Various methods exist concerning the extrapolation of wind speed to the hub height of a wind turbine. The wind velocity varies as a disproportionate function of height. At low height levels surface friction and low level obstacles introduce turbulence and reduce the observed wind velocity. The velocity $u(z)$ at low height levels over consistent terrain or see surfaces is conventionally approximated by the logarithmic function, of u^* is the scaling velocity, k is the Von Karman's constant, usually equal to 0.4, z is the desired height level and z_0 the roughness length, values lower than 0.01, while the roughness length or surface friction is the parameter of most influence in equation, defined as:

$$u(z) = \frac{u^*}{k} \ln\left(\frac{z}{z_0}\right) \tag{4}$$

There are several theoretical expressions used for determining the wind speed profile. However, the increase of wind speed with height should be considered for the installation of large wind turbines. Thus, the surveys must rely to simpler expressions and secure satisfactory results even when they are not theoretically accurate. For h_0 = 10 m and z_0 = 0.01 m, the parameter α = 1/7, which is consistent with the value of 0.147 used in the wind turbine design standards (IEC standard, 61400-3, 2005) to represent the change of wind speeds in the lowest levels of the atmosphere. Wind speed is usually recorded at the standard meteorological height of 10 m, while wind turbines usually have hub heights near 80 m. In cases which lack of elevated measurements, the hub-height wind velocity is estimated by applying a vertical extrapolation to the surface or reference measurements. However, the vertical extrapolation coefficient may contain errors and uncertainties due to terrain complexity, ground or see conditions, atmospheric stability, and turbulence [2–4, 17–36]. The wind speed $v(z)$ at a height z can be calculated directly from the wind speed $v(z_{ref})$ at height zref (usually the standard measurement level) by using the logarithmic law (the so-called Hellmann exponential law) expressed by:

$$\frac{v(z)}{v_0} = \left(\frac{z}{z_{ref}}\right)^{\alpha} \tag{5}$$

where, $v(z)$ is the wind speed at height z, v_0 is the speed at z_{ref} (usually 10 m height, the standard meteorological wind measurement level), and α is the power law index. This coefficient is a function of the site surface roughness and the thermal stability, frequently assumed to be 1/7 for open land. However, this parameter can vary diurnally, seasonally and spatially. It was found that a single power law is insufficient to adequately estimate the wind power at a given site, especially during nighttime and in presence of the low-level jets. Another formula, the logarithmic wind profile law, widely used in Europe, is:

$$\frac{v}{v_0} = \frac{\ln(z/z_0)}{\ln(z_{ref}/z_0)} \tag{6}$$

Here, z_0 is again the roughness length, expressed in meters, depending basically on the surface type, ranging from 0.0002 up to 1.6 or higher. In addition to the roughness, these values can vary during the day and at night, and even during the year. Once wind speeds have been calculated at other heights, Eq. (6) can be used for calculating the useful wind energy potential. Notice that the wind shear over the rotor area can also be significant. The standard procedure for power curve measurements is given by the IEC Standard, 6-1400-12-1, 2005 [19], where the wind speed at hub height is considered representative of the wind over the rotor area. This assumption can lead to large wind power estimate inaccuracies since inflow is often non-uniform and unsteady over the rotor-swept area. By integrating the wind profile over the rotor span, a corrected wind speed is obtained:

$$U_{avrg} = \frac{1}{D}\int_{H-\frac{D}{2}}^{H+\frac{D}{2}} v(z)dz = v(H)\cdot\frac{1}{\alpha+1}\cdot\left(\left(\frac{3}{2}\right)^{\alpha+1} - \left(\frac{1}{2}\right)^{\alpha+1}\right) \tag{7}$$

where H is the nacelle height and D is the rotor diameter. It is observed these corrections have less significant effects. For wind speeds in the range 4 m/s to 20 m/s (the useful wind speed regime) the corrected power differs less than 5% from the uncorrected power. However, the corrected power is larger than the uncorrected turbine power.

An additional wind velocity property that can make the impact on wind turbine operations is the wind gustiness. Proper wind turbine design and operation requires knowledge of wind extremes and gustiness, defined by the wind gust factor. This is important in areas where wind climate shows strong gusty winds, e.g. downslope windstorms [3, 50, 51]. In sites with higher turbulent intensity and gusty winds, turbines are subject to extreme structural loading and fatigue. The gust factor (G) is defined as:

$$G = \frac{u_g}{U} - 1 \tag{8}$$

where ug is the gust speed and U is the mean daily wind speed. Higher gusts are usually associated with higher mean wind speeds; however, it also is expected that normalized gust speed ug/U and, consequently, the gust factor, G, decreases with the increasing mean speed. The following equation relates the gust factor to the mean daily wind speed:

$$G = AU^n \tag{9}$$

where the parameters A and n are obtained by using a least-square fit of the logarithm of G vs. the logarithm of the mean daily wind speed. While gusts generally decrease as wind speed increases, in extreme cases the wind gusts can easily reach over twice the strongest wind speeds ($v > 20\ ms^{-1}$) and damage a wind turbine. However, wind gusts over 25 m/s, the upper wind speed limit of a large wind turbine, are quite unlikely in many areas, occurring only about 2% of time [2–4, 23–25]. Gusts associated with stronger winds may cause considerable losses by reducing the energy production of the wind turbine which would otherwise operate at nominal output power. Another effect of wind gust is the additional stress on the wind turbine structure, which may reduce its lifespan.

The low-level jet, observed worldwide is a mesoscale phenomenon associated with the nighttime very stable boundary layer that can have a width of hundreds of kilometers and a length of a thousand kilometers. During nighttime and over land, ground surface cools at a faster rate than the adjacent air and stable stratification forms near the surface and propagates upward. Downward mixing of the winds is reduced, and winds aloft become decoupled from the surface and accelerate. The maximum wind speeds are usually 10 m/s to 20 m/s or even higher at elevations 100 m to 300 m. Consequently, it is not possible to accurately estimate winds aloft at hub and blade tip heights from routine surface measurements [4]. Additionally, a strong wind shear and associated turbulence are developing at the bottom and top of the jet layer. After the wind velocity field data are collected and transferred to the computing environment, the next steps are to validate and process data, and generate reports. Data validation is defined as the inspection of all the collected data for completeness and reasonableness, and the elimination of erroneous values. Data validation transforms raw data into validated data. The validated data are then processed to produce the summary reports required for analysis, step crucial in maintaining high rates of data completeness. Therefore, data must be validated as soon as possible, after they are transferred. The sooner the site operator is notified of a potential measurement problem, the lower the risk of data loss. Data can be validated either manually or by using computer-based data analysis [2–4]. The latter is preferred to take advantage of the computer power and speed, although some manual reviews are always required. Data validation implies visual inspection, editing, missing data interpolation, outliers and questionable data rejection, and finally saving data in appropriate format.

3. Wind energy statistical analysis

The wind is characterized by its speed and direction which is affected by several factors, including: geographic location, climate characteristics, height above ground, vegetation and surface topography. Wind turbines interact with the wind capturing part of its kinetic energy and converting it into usable energy. Wind availability, the influence of the turbine height installation above ground, the wind gusting effect and the wind turbine micro-sitting are the main influences of the annual energy output and are the theoretical basis for the wind energy assessment [2–9, 14, 32–60]. There main aspects of the wind resource assessment are the wind power potential estimate, and the prediction of the wind plant energy production. The measured wind velocity data were usually available at 10 m standard meteorological height. However, sometimes the anemometers are installed on top of buildings or airport control towers, or at meteorological masts. The wind energy classes were developed for 10 m height because that was the standard for meteorological data, and then the wind power potential is extrapolated at 50 m, assuming that the wind shear exponent was 1/7 for all locations. Global wind patterns, upper air wind data, and boundary layer meteorology were routinely used to obtain estimates of the wind energy resource [2–4]. The knowledge of the quasi-steady mean wind speeds that can be expected at a potential site is critical to determine the wind economic viability of a wind energy project. Such information is essential in the wind turbine selection in order to maximize efficiency and durability. The wind frequency distributions, used in wind energy assessment are predicted from wind measurements collected during several years. If such information is not available, wind speed probability distributions, constructed from limited field campaigns, e.g. Weibull or Rayleigh probability distributions are used to estimate the wind power potential. The highly dependent nature of energy production on wind speed needs accurate predictions of the distribution of wind speeds for a prospecting wind farm location and for accurate energy production calculations.

The wind probability distribution (PDF) functions have been investigated, employed and explained by many researchers and engineers involved in the wind energy [2–12, 42–68]. In wind power studies such probability distributions are used for assessment and analysis of wind energy resources, wind power plant operation, as well as for turbine design. Both analytical and numerical methods can be carried out. However, a planning from a different point of view can be performed, while similar distribution functions can be used for wind power, if wind velocity distribution functions are taken into account, together with WT features, provided by the manufacturers. Usually the wind velocity time series are rather large, differences among parameter estimation methods is not as important as differences among distributions. There are several estimators of PDF parameters, such as Method of Moment (MOM), Maximum Likelihood Estimators (MLE), Least-Square (LS), and Percentile Estimators Methods [2–72]. These estimators are unbiased, so there is no reason to give preference to any of them. The choice of specific estimators is based on the existing wind speed data, computing availability and user preference. The rule-of-thumb is to select a number of estimators of the PDF parameters, while the parameters are usually computed by taking the averages of the estimates found by these methods. We preferentially use MLE, MOM and LS estimators for the large samples, and the averages are the PDF parameters [45–70]. However, when using MOM, we calculate the sample mean standard deviation (s), and skewness (G) as:

$$\overline{v} = \frac{1}{N}\sum_{i=1}^{N} v_i \tag{10}$$

$$s = \sqrt{\frac{1}{N-1}\sum\nolimits_{i=1}^{N}(v_i - \overline{v})^2} \tag{11}$$

and

$$G = \frac{1}{N} \cdot \frac{\sum_{i=1}^{N}(v_i - \overline{v})^3}{s^3} \tag{12}$$

where N is the number of observations in v, the wind speed. Once the wind power distribution function is obtained, the mean power available is deduced. So as not to depend on the type of wind turbine, this will be shown per unit of surface (mean power density). This process is performed in four different ways: (1) obtaining of the wind power; (2) Betz' law considerations; (3) consideration of realistic values, remembering that Betz' law is an upper limit; and (4) consideration of WT parameters such as Cut-In and Cut-Out wind speed, rated speed, and rated power. The goal of any wind energy assessment and analysis is to give response to questions about statistical distribution of the maximum power obtainable from the wind, regardless of the WT chosen, and also taking into account its features, when the only input value is the mean wind speed.

3.1 Weibull probability distribution

The Weibull density distribution is a commonly applied statistical distribution to model wind speed regime. The use of probability distribution functions in order to define and to characterize the field data has a long history of use. It has been established, in the literature [2–4, 41–70] that the Weibull probability distribution is very well fitted to characterize wind speed regimes, being commonly used to estimate and to assess wind energy potential. However, efforts have been made over the years to fit the wind data to other distributions, e.g. exponential distribution, gamma distribution, or logistic distribution. The Weibull probability distribution is given by:

$$f_{WB} = k\frac{v^{k-1}}{c^k}\exp\left(-\left(\frac{v}{c}\right)^k\right) \tag{13}$$

The Weibull distribution is a function of two parameters: k, the shape parameter, and c, the scale factor. These parameters are defining the shape or steepness of the curve and the mean value of the distribution. These coefficients are adjusted to match the wind data at a particular site. For wind modeling, typical k values range from 1 to 2.5 and can vary drastically form site to site, as well as during years and/or seasons. The scale parameter, c, corresponds to the average wind speed for the site. The main inaccuracy of the Weibull distribution is that it always has a zero probability of zero wind speed, which is not the case since there are frequently times in which no wind is blowing. The higher the k value, the sharper the increasing part of the curve is. The higher c values correspond to a shorter and fatter distribution, with a higher mean value. Ideally the mean value would correlate with the rated wind speed of the turbine: producing rated power for the extended time annually. The cumulative probability function for Weibull distribution is given by:

$$F(v) = 1 - \exp\left[-\left(\frac{v}{c}\right)^k\right] \tag{14}$$

The availability of high quality wind speed distributions is crucial to accurate forecasts of annual energy production for a wind turbine. Statistical distributions suffice for early estimations, while the actual wind speed measurements are necessary for accurate predictions. The factors k and c featuring in Eq. (13) are determined for each measurement site. There are several estimators of Weibull parameters, such as the Moment (MOM), Maximum Likelihood (MLE), Least-Square, and Percentile Estimators. These estimators are unbiased, although some of them, such as the Method of Moments, may have large variances, so there is no reason to prefer any of them. The choice of specific estimators is based on the existing wind speed observations, computing availability and user preference. The rule-of-thumb is to select estimators of the Weibull parameters, such as: standard least-square, maximum likelihood or MLE variants, while the shape and scale parameters are computed taking the averages [2–4]. If sufficient wind speed observations are available, one of the most used is the MOM methods or its variants. It is based on the numerical iteration of the following two equations while the mean ($\overline{v}$) and standard deviation (s) of the wind speeds are determined from:

$$\overline{v} = c \cdot \Gamma\left(1 + \frac{1}{k}\right) \tag{15}$$

and

$$s = c\left[\Gamma\left(1 + \frac{2}{k}\right) - \Gamma^2\left(1 + \frac{1}{k}\right)\right]^{1/2} \tag{16}$$

where $\overline{v}$ is the wind speed data set (sample) mean, as defined in Eq. (11), s is the wind speed data set (sample) standard deviation, and $\Gamma()$ is the Gamma function:

$$\Gamma(x) = \int_0^\infty t^{x-1} \exp(-t) dt \tag{17}$$

A special case of the moment method is the so-called empirical method, where the Weibull shape parameter *k* is estimated by following relationship:

$$k = \left(\frac{s}{\overline{v}}\right)^{-1.086} \tag{18}$$

Then the scale parameter, c is computed by using the following relationship:

$$c = \frac{\overline{v}}{\Gamma(1 + 1/k)} \tag{19}$$

Both, the moment and empirical method require a reasonable wind speed observations data set to be available. Another Weibull parameter estimator, based on the least squares, is the graphical method. In which a straight line is fitted to the wind speed data using lease squares, where the time-series data must be sorted into bins. Taking a double logarithmic transformation, the cumulative distribution function is rewritten as:

$$ln\{-ln[1 - F(v)]\} = kln(v) - kln(c)$$

Plotting ln(v) against ln{−ln[1 - F(v)]}, the slope of the best fitted line to data pairs is the shape parameter, and the scale parameter is then obtained by the intercept with y-axis. The graphical method requires that the wind speed data be in cumulative frequency distribution format. Time-series data must therefore first sort into bins. In essence the graphical method is a variant of the moment method, consisting of the calculation from the time series of the observations of statistical estimators, such as wind speed means and of the square wind speeds. Then the Weibull parameters are calculated as:

$$\overline{v} = \frac{c}{k}\Gamma\left(\frac{1}{k}\right) \tag{20}$$

and

$$\overline{v^2} = \frac{2c^2}{k}\Gamma\left(\frac{2}{k}\right) \tag{21}$$

Then computing the means from the wind observations (Eq. (20) and Eq. (21)), the Weibull parameters are estimated. The Weibull distribution can also be fitted to time-series wind data using the maximum likelihood method. The shape factor k and the scale factor c are estimated, numerically using the following two equations:

$$k = \left(\frac{\sum_{i=1}^{N} v_i^k \, ln(v_i)}{\sum_{i=1}^{N} v_i^k} - \frac{\sum_{i=1}^{N} ln(v_i)}{N}\right)^{-1} \tag{22}$$

$$c = \left(\frac{1}{N}\sum_{i=1}^{N} v_i^k P(v_i)\right)^{1/k} \tag{23}$$

where v_i is the wind speed in the bin i and N is the number of nonzero wind speeds (the actual wind speed observations). Eq. (22) is solved numerical, usually through iterative methods, with k equal to 2 as initial guess, and then Eq. (23) is solved explicitly. When wind speed data are available in frequency distribution format, a MLE variant can be applied. In this cans, the Weibull parameters are then estimated through:

$$k = \left(\frac{\sum_{i=1}^{N} v_i^k \, ln(v_i) P(v_i)}{\sum_{i=1}^{N} v_i^k P(v_i)} - \frac{\sum_{i=1}^{N} ln(v_i) P(v_i)}{P(v \geq 0)}\right)^{-1} \tag{24}$$

$$c = \left(\frac{1}{P(v \geq 0)} \sum_{i=1}^{N} v_i^k P(v_i) \right)^{1/k} \tag{25}$$

where, $P(v_i)$ is the frequency with which the wind speed falls within bin i, $P(v \geq 0)$ is the probability that the wind speed equals or exceeds zero. Eq. (23) must be solved iteratively, after which Eq. (25) is solved explicitly to determine the Weibull parameters. One of the parameter estimator not very often used is the energy pattern factor method. In this approach, the energy pattern factor for a given wind speed data is defined as:

$$E_{pf} = \frac{\overline{v^3}}{\overline{v}^3} \tag{26}$$

here $(\overline{v^3})$ is the mean of the cubes of the wind speed. Notice that the factors in Eq. (26) are related to the wind energy estimates. Weibull shape parameter can be estimated with the following equation:

$$k = 1 + \frac{3.69}{E_{pf}} \tag{27}$$

Scale parameter is estimated by using Eq. (19), for example. Often isw necessary to estimate the Weibull parameters in the absence of suitable information about the distribution of wind speeds. For example, if only annual or monthly averages may be available, the value of k must be estimated by Eq. (27). The value of k is usually from 1.5 and 3, depending on the wind variability. Smaller k values correspond to variable winds.

To analyze the accuracy of the aforementioned methods, the following tests are used, RMSE (root mean square error), χ^2 (chi-square), R^2 (variance analysis or method efficiency) and the Kolmogorov–Smirnov test. These tests are examining whether a PDF is suitable to describe the wind speed data or not. The RMSE test is defined by:

$$RMSE = \left[\frac{1}{N} \sum_{i=1}^{N} (y_i - x_i)^2 \right]^{1/2} \tag{28}$$

where y_i is the actual values at time stage i, x_i is the value computed from correlation for the same stage, and N is the number of data. The next two tests are defined by:

$$\chi^2 = \frac{\sum_{i=1}^{N} (y_i - x_i)^2}{N - n} \tag{29}$$

and

$$R^2 = \frac{\sum_{i=1}^{N} (y_i - z_i)^2 - \sum_{i=1}^{N} (y_i - x_i)^2}{\sum_{i=1}^{N} (y_i - z_i)^2} \tag{30}$$

where N is the number of observations, y_i is the frequency of observations, x_i is the Weibull frequency, and z_i is the mean wind speed. The Kolmogorov–Smirnov test is defined as the max-error between two cumulative distribution functions:

$$Q = max\,|F_T(v) - F_O(v)| \tag{31}$$

where $F_T(v)$ and $F_O(v)$ are the cumulative distributions functions for wind speed not exceeding v computed by using estimated Weibull parameters and by observed (or randomly generated) time-series, respectively. The critical value for the Kolmogorov–Smirnov test at 95% confident level is given by:

$$Q_{95} = \frac{1.36}{\sqrt{N}} \tag{32}$$

If Q value exceeds the critical value, then there is significant difference between the theoretical and the time-series data under the given confident level. **Figure 1** shows the fitted Weibull probability distributions with long-term observations, for two locations [4, 51, 52], exhibiting a good agreement between fitted PDF and actual data. The energy that a wind turbine generates depends on both on its power curve, a nonlinear relationship between the wind speed and turbine power output and the wind speed frequency distribution. If derived from long-term (multi-annual) wind speed data sets the histograms shape of the PDFs that are characterizing the wind speed at a specific site or for a region.

3.2 Other probability distribution function used in wind energy

Another used probability distribution is the Rayleigh distribution, which is a special case of the Weibull distribution where k = 2. The Rayleigh distribution is simpler because it depends only on the mean wind speed, and is given by:

$$f_{RL}(v) = \frac{\pi}{2}\frac{v}{c^2} exp\left[-\frac{\pi}{4}\left(\frac{v}{c}\right)^2\right] \tag{33}$$

These two probability distribution functions, Weibull and Rayleigh [2–4, 9, 41–72] are the most commonly used for wind energy analysis and assessment. The Rayleigh PDF is a special case of the Weibull distribution with k = 2. Notice that for both distributions, $V_{min} = 0$ and $V_{max} = \infty$. The cumulative Rayleigh distribution is expressed as:

$$F(v) = 1 - e^{-(v/c)^2} \tag{34}$$

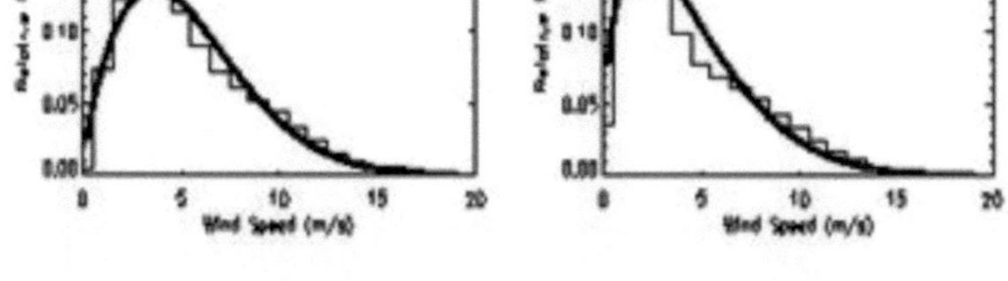

Figure 1.
Experimental wind speed probability density functions, at 50 m level, using the composite 2003–2008 data sets, for two 50 m instrumented towers [4, 51, 52].

For the Rayleigh distribution the single parameter, c, relates the following three properties:

$$c = V_{mp}\sqrt{2} = \frac{2\mu}{\sqrt{\pi}} = \sigma\sqrt{\frac{4}{8-\pi}} \tag{35}$$

The Rayleigh distribution can be written using V_{mp} or the mean velocity, μ. Determination of the mean and standard deviation from experimental data for the normal distribution are well known. The MLE estimate of the normal distribution en is the arithmetic mean. The parameter c in the Rayleigh distribution can be evaluated from N wind velocities, V_i. If experimental data are used to determine distribution parameters, the computed result is called an estimate of the true parameter. Here, the symbol used to indicate that the equation gives only an estimate of the true distribution parameter, c.

$$\hat{c} = \sqrt{\frac{1}{2N}\sum_{i=1}^{N} V_i^2} \tag{36}$$

The 3-parameter Weibull (W3) is a generalization of the 2-paramter Weibull distribution, where the location parameter s establishes a lower bound (assumed to be zero in the case of 2-parameter Weibull distribution). It was found that for some areas the W3 fits wind speed data better than the 2-parameter Weibull model [4, 40–72]. The W3 probability distribution and cumulative distribution functions are expressed as:

$$f(v,k,c,\tau) = \frac{kv^{k-1}}{c^k}\exp\left[-\left(\frac{v-\tau}{c}\right)^k\right] \tag{37}$$

and

$$F(v,k,c,\tau) = 1 - \exp\left[-\left(\frac{v-\tau}{c}\right)^k\right] \tag{38}$$

Respectively, for $v \geq \tau$, and τ is the location parameter, locating the probability distribution along the abscissa (v axis). Changing the value of τ has the effect of sliding the distribution and its associated function either to the right (if $\tau > 0$) or to the left (if $\tau < 0$). The MLE estimators for the W3 PDF parameter calculation are:

$$\frac{\sum_{i=1}^{N}(v_i-\hat{\tau})^{\hat{k}}\ln(v-\hat{\tau})}{\sum_{i=1}^{N}(v_i-\hat{\tau})^{\hat{k}}} - \frac{1}{\hat{k}} - \frac{1}{N}\sum_{i=1}^{N}\ln(v_i-\hat{\tau}) = 0 \tag{39}$$

and

$$\hat{c} = \left(\frac{1}{N}\sum_{i=1}^{N}(v_i-\hat{\tau})^{\hat{k}}\right) \tag{40}$$

$$\hat{\tau} + \frac{\hat{c}}{N^{1/\hat{k}}}\Gamma\left(1+\frac{1}{\hat{k}}\right) = U_{\min} \tag{41}$$

where N is the number of observations in the sample v, and U_{min} indicates the minimum values in the v time series. Parameters of the W3 distributions are then found iteratively.

Another PDF used in wind energy assessment, especially in offshore applications is the lognormal distribution [4, 66–68]. The 2-parameter lognormal PDF is given by:

$$\widehat{\tau} + \frac{\widehat{c}}{N^{1/\widehat{k}}}\Gamma\left(1 + \frac{1}{\widehat{k}}\right) = U_{\min} \tag{42}$$

And its cumulative distribution function is expressed as:

$$F(v, \mu, \sigma) = \frac{1}{2} + \frac{1}{2}erf\left[\frac{\ln(v) - \mu}{\sigma\sqrt{2}}\right] \tag{43}$$

where $erf(x) = \frac{2}{\sqrt{\pi}}\int_0^x \exp(-t^2) \cdot dt$ is the error function from the Normal distribution, and the parameters μ and σ are the mean and standard deviation of the natural logarithm of v. The parameter estimators are given by:

$$\widehat{\mu} = \ln\left(\frac{\overline{v}}{\sqrt{1 + \frac{s^2}{\overline{v}^2}}}\right)$$

$$\widehat{\sigma} = \sqrt{\ln\left(\frac{s^2}{\overline{v}^2}\right)}$$

The truncated normal distribution is the probability distribution function of a normally distributed random variable whose values are either bounded at lower end, higher end or at both. Since the wind speed is only positive, the most common is the single truncated normal distribution, suitable for nonnegative case:

$$n(v, \mu, \sigma) = \frac{1}{I(\mu, \sigma)\sigma\sqrt{2\pi}} \exp\left[-\frac{(v - \mu)^2}{2\sigma^2}\right], \; for\; v > 0 \tag{44}$$

where μ and σ are the date mean and standard deviation, and I(μ, σ) is the normalized factor, making the integral of this distribution equal to one, the cumulative distribution function is evaluated in its domain de definition. The normalized factor is given by:

$$I(\mu, \sigma) = \frac{1}{\sigma\sqrt{2\pi}}\int_0^\infty \exp\left[-\frac{(v - \mu)^2}{2\sigma^2}\right]dv \tag{45}$$

The distribution function parameters can be determine using graphical, moment or maximum likelihood methods or a combination of them. The maximum entropy probability (MEP) concept has commonly been applied in many engineering and science areas. The entropy of PDF, f(x) is defined [54–72] by:

$$S(x) = -\int f(x)\ln(f(x))dx$$

Maximizing the entropy subject to specific constrains enables to find the most likely probability distribution function if the information available is provided by moment functions. The classical MEP solution applied to wind distribution case is given by:

$$f(v) = \exp\left[-\sum_{k=0}^{N} a_k v^k\right] \tag{46}$$

and solution, Lagrange multipliers are given by the following nonlinear system of equations:

$$Z_n(a) = \int v^n \exp\left[-\sum_{k=0}^{n} a_k v^k\right] dv = \lambda_n \, for \; n = 1, 2, \ldots, N$$

The λ_n, n = 0,1, ..., N, with $\lambda_0 = 1$, are the distribution moments, representing the mean values of n wind speed powers, calculated from observations. MEP probability density functions of third or fourth order with three or four moment constrains.

Gamma PDF can be expressed with the following function:

$$g(v; x; \beta) = \frac{v^{x-1}}{\beta^2 \Gamma(x)} \exp\left[-\frac{v}{\beta}\right], \; for \; v, x, \beta > 0 \tag{47}$$

where x and β are the shape and scale parameter, respectively. The parameters of the Gamma distribution can be estimated using graphical, moment or maximum likelihood methods, similar to one presented above in the Weibull case. However, the Gamma PDF is usually employed in a mixture of distributions in connection with Weibull PDF.

In recent years, in order to improve the accuracy of wind statistics, mixtures of PDFs were employed [4, 50–72]. Distribution function mixed with Gamma, Weibull, or Normal distribution functions can be used to describe the wind statistics. For example, Gamma and Weibull mixture applied to wind energy assessment is given by:

$$h(v; w; x, \beta, k, c) = wg(v, x, \beta) + (1 - w)f(v, k, c) \tag{48}$$

where $0 \le w \le 1$ is the weight parameter indicating the mixed proportion of each probability distribution included in the PDF mixture relationship. Again, the five parameters, in the Eq. (48) can be estimated using graphical, moment, maximum likelihood methods or any combination them, as discussed in the Weibull case.

3.3 Wind turbine power and energy estimates

The average power of a wind turbine, over a specific time period (e.g. one month) is determined by multiplying the wind speed probability density function $f_{PDF}(v)$ and the power curve of the turbine, $CP(v)$. The power curve represents the wind turbine output power vs. wind speed diagram [2–4, 68–72]. The power curve depends on the wind speed, air density and turbulence intensity, being usually provided by the manufacturers or can be constructed from the field measurements. However, the manufacturers do not usually are providing the information on the power curves of their wind turbines in a continuous (analytic) form, but rather in a discrete form with N nodes (P_{WT-i}, v_i). Wind power density, a measure of the energy flux through an area perpendicular to the direction of motion, it is extensively used in wind energy assessment. It is also varying not only with the cube of wind speed, but also with the air density and turbulence. The monthly average energy produced by a wind turbine is obtained from the monthly average output power by multiplying it with the hours of that month. The prediction of the wind speed variation with height, the variation in wind speed over the wind farm area, air density and the wake interaction between the wind turbines are usually calculated by using a computer programs, specifically designed to facilitate accurate predictions of wind farm energy production. Such computing applications allow fast computations of the energy production for different layouts, turbine type and hub height to determine the optimum setting. In the statistic approaches, based on an estimate of the wind velocity probability distribution function, for the location wind regime, $f_{PDF}(v)$, extrapolated at the hub height and a known WT power curve $P_{WT}(v)$, the mean power output of a wind turbine (assuming 100% of its availability) is given in the following equation:

$$\overline{P}_{WT} = \int_{V_{ci}}^{V_{co}} f_{PDF}(v) \cdot CP(v) \cdot dv \tag{49}$$

Generally, the integral of Eq. (49) has no analytic solution and must be resolved numerically. For example, if the Weibull probability distribution is determined the wind power density, *WPD* is estimated, function of the Weibull parameters by:

$$WPD = 0.5 \cdot \rho c^3 \left(1 + \frac{3}{k}\right)$$

The relationship between the power output of a turbine and the incoming wind speed is usually simplified by a generic power curve model, as expressed in practical application by the following relationship:

$$P_{WTG}(v) = \begin{cases} 0, & \mathrm{v} \leq \mathrm{V_{ci}} \text{ or } \mathrm{V_{co}} \leq \mathrm{v} \\ P_{WT-Rated}\left(\frac{v}{V_R}\right)^3, & \mathrm{V_{ci}} \leq \mathrm{v} \leq V_R \\ P_{\text{WT-Rated}}, & V_R \leq \mathrm{v} \leq \mathrm{V_{co}} \end{cases} \tag{50}$$

Here, $P_{WT\text{-}Rated}$ is the wind turbine rated power, V_{ci}, V_R, and V_{co} are the wind turbine cut-in, rated, and cut-off speeds. These values are also provided by the manufacturer. The total power yield of the wind farm is the sum of the power output of each wind farm turbine. For a number of wind turbines taking into consideration of the generator efficiency, the total output power can be extracted as follows:

$$P_{WT-tot} = N_{WT} \times \eta_{WTG} \times P_{WT} \tag{51}$$

Where: η_{WTG} and N_{WT} are the wind generator efficiency and number of wind turbine generators in the wind farm, respectively. Wind turbine output power is estimated, as adapted from the Eq. (49), by a relationship such as:

$$P_{WT}(v) = \begin{cases} 0, & \mathrm{v} \leq \mathrm{V_{ci}} \text{ or } \mathrm{V_{co}} \leq \mathrm{v} \\ 0.5 \cdot \rho \cdot \mathrm{A} \cdot \mathrm{C}_P(\lambda, \beta) v^3, & \mathrm{V_{ci}} \leq \mathrm{v} \leq V_R \\ P_{\text{WT-Rated}}, & V_R \leq \mathrm{v} \leq \mathrm{V_{co}} \end{cases} \tag{52}$$

The wind generator power output is best estimated through interpolation of the values of the data provided by the manufacturers with wind velocity observations. As the power curves are quite smooth, they can be approximated using a cubic spline interpolation. The fitting equation of the output characteristic of wind generator can be expressed as:

$$P_{WTG}(v) = \begin{cases} 0 & v \le V_{ci} \quad or \quad v \ge V_{co} \\ a_1 v^3 + b_1 v^2 + c_1 v + d_1 & V_{ci} < v < V_1 \\ a_2 v^3 + b_2 v^2 + c_2 v + d_2 & V_1 < v < V_2 \\ \cdots & \\ a_n v^3 + b_n v^2 + c_n v + d_n & V_{n-1} < v < V_r \\ P_{rated} & V_r \le v < V_{co} \end{cases} \tag{53}$$

Where P_{WTG} is the output power of wind generator at wind speed v, is the wind turbine rated power; v is the wind speed at the hub height, n is the number of cubic spline interpolation functions corresponding to $n + 1$ values, couples of the wind speed and the wind turbine output power. These data are provided by the manufacturers, a, b, c and d are the polynomial coefficients of the cubic spline interpolation functions, depending on the wind turbine type and are estimated from measurements. Simplified versions of the wind turbine power output calculations, usually used in practice and design are:

$$P_{WT} = \begin{cases} P_{WT-Rated} \dfrac{(v(t) - V_{ci})}{(V_r - V_{ci})}, & for \ V_{ci} \le v(t) < V_R \\ P_{WT-Rated}, & for \ V_R \le v(t) \le V_{co} \\ 0, & for \ v(t) < V_{ci}, or \ v(t) > V_{co} \end{cases}$$

Or

$$P_{WT} = \begin{cases} a \cdot (v(t))^3 - b \cdot P_{WT-Rated}, & for \ V_{ci} \le v(t) < V_R \\ P_{WT-Rated}, & for \ V_R \le v(t) \le V_{co} \\ 0, & for \ v(t) < V_{ci}, or \ v(t) > V_{co} \end{cases}$$

where $P_{WT}(v)$ is the output power of WT units with an ambient wind speed v, $P_{WT\text{-}Rated}$ is the rated power of the WT, Vci, Vco and VR are the cut-in wind speed, cut-out wind speed, and rated wind speed, respectively. Based on the wind speed data, the output power of WT units is calculated according to one of these relationships. The parameters a, and b, of the last WT power output relationship are calculated by these expressions:

$$a = \frac{P_{WT-rated}}{V_r^3 - V_{ci}^3}, \text{ and } b = \frac{V_{ci}^3}{V_r^3 - V_{ci}^3}$$

Calculation of energy output requires knowledge of the wind speed probability distribution and the wind turbine power curve. Once, the wind speed probability distribution and turbine power curve are determined, the output energy of a wind turbine or for a wind farm can be easy determined. The wind energy (EWT) that can be extracted by a wind turbine, over a *T* time period is defined by this relationship:

$$E_{WTG} = \int_0^{\infty} CP(v) \cdot f_{PDF}(v) \cdot dv \tag{54}$$

3.4 Wind direction

To ensure the most effective use of a wind turbine it should be exposed to the most energetic wind. Though the wind may blow more frequently from the west more wind energy may come from a different direction if those winds are stronger. It is important to find out which directions have the best winds for electricity production. The distribution of wind direction is crucially important for the evaluation of the possibilities of utilizing wind power. The distributions of wind speed and direction are conventionally given by wind roses. A wind rose, generated from your wind resource assessment, is a helpful tool to determine wind direction and distribution. Traditionally, wind direction changes are illustrated by a graph, which indicates percent of winds from that direction, or the wind rose diagram [2, 4, 52, 61, 62, 66–70]. The wind rose diagrams and wind direction frequency histograms provide useful information on the prevailing wind direction and availability in different wind speed bin. A vane points toward the wind source. Wind direction is reported as the direction from which the wind blows, not the direction toward which the wind blows, e.g. a North wind blows from the North toward the South. The wind direction varies from station to station due to different local features (topography, altitude, distance from the shore, vegetation, etc.). There are usually changes in the wind directions on diurnal, seasonal or annual basis. The wind direction can also be analyzed using continuous probability models to represent distributions of directional winds, e.g. von Mises circular statistics, usually comprised of a mixture of von Mises distributions.

The wind rose diagrams and wind direction frequency histograms provide useful information on the prevailing wind direction and availability of directional wind speed in different wind speed bins. The wind roses were constructed using the composite data sets of measurements of wind velocities. The wind direction is analyzed using a continuous variable probability model to represent distributions of directional winds, comprising of a finite mixture of the von Mises distributions (vM – PDF) [2–4, 52, 61, 62, 60–66]. The model parameters are estimated using the least square method. The range of integration to compute the mean angle and standard deviation of the wind direction is adjusted to minimum variance requirements. For example, the proposed probability model mvM(θ) is comprised of a sum of N von Mises probability density functions, vMj(θ), as:

$$mvM(\theta) = \sum_{j=1}^{N} w_j vM_j(\theta) \tag{55}$$

where w_j are non-negative weighting factors that sum to one [2–4, 52, 61, 62, 66]:

$$0 \le w_j \le 1, \quad \text{for } j = 1, \ldots, N, \text{and } \sum_{j=1}^{N} w_j = 1$$

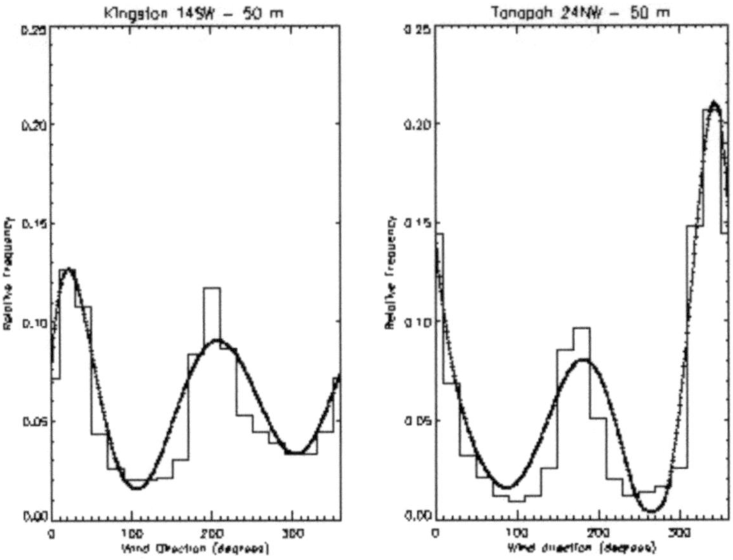

Figure 2.
Frequency histograms of wind directions and the fitted von Mises distribution functions at (a) Kingston 14SW tower and 50 m level, and (b) Tonopah 24NW and 50 m level using the composite 2003–2008 data sets [4, 51, 52].

A von Mises distribution vM-PDF if its probability is defined by the equation:

$$vM_j\left(\theta; k_j, \mu_j\right) = \frac{1}{2\pi I_0\left(k_j\right)} \exp\left[k_j \cos\left(\theta - \mu_j\right)\right], \text{and } 0 \leq \theta \leq 2\pi \tag{56}$$

where $k_j \geq 0$ and $0 \leq \mu_j \leq 2\pi$ are the concentration and mean direction parameters. The angle corresponding to the northerly direction is taken as 0°. Note that in meteorology, the angle is measured clockwise from North. Here, $I_0(k_j)$ is a modified Bessel function of the first kind and order zero and is given by:

$$I_0\left(k_j\right) = \frac{1}{2\sqrt{\pi}} \int_0^{2\pi} \exp\left[k_j \cos\theta\right] d\theta \approx \sum_{p=0}^{\infty} \frac{1}{(p!)^2} \left(\frac{k_j}{2}\right)^{2p} \tag{57}$$

The distribution law mvM(θ), given by Eq. (49) is numerically integrated between two given values of θ to obtain the probability that wind direction is within a particular angle sector. Various methods are employed to compute the 3 N parameters on of the mixture of von Mises distribution. **Figure 2** is showing fitted von Mise distributions to two long-term wind direction time series for two locations [4, 52, 53, 66].

4. Measure-correlate-predict methods

Measure-correlate-predict (MCP) algorithms are used to predict the wind energy resource at target sites, by using relatively long-term measurements at reference locations. MCP methods model the relationship between wind velocity data measured at the target site, usually over a short period, up to a year, and concurrent data at a nearby reference site. The model is then used with long-term data from the reference site to predict the long-term wind speed and direction distributions at the target sites [4, 73–76]. Typical wind energy assessments last anywhere from one to three years, with important decisions to be taken often after only few months, there is an obvious need for a prediction of the performance of a planned wind energy project for expected life time (20 years or more). Such an assessment is an important part of the financing process. While the measurement campaign may correspond to an untypically high or low period, correlations with nearby reference stations help to detect such trends and provide accurate long-term estimate of the wind velocity at the development site and its inter-annual variations. Moreover, since the wind turbine power output depends on the wind speed in a non-linear way, the distribution of the wind speeds should be predicted accurately. MCP methods proceed by measuring the winds at a target site, correlating with winds from reference sites, then by applying these correlations to historical data from the reference site, to predict the long term wind resource of the target site [2–4, 72–77].

Several MCP algorithms have been studied using wind data from potential wind energy sites. Some of the algorithms and methods have been improved using probabilistic methods, and have then been implemented into software packages, such as the WindPRO or WAsP for planning and projecting of wind power plants. MCP process methods [72–77] consist of: (1) collect wind data at the predictor site for extended possible time period; (2) identify reference sites, for which high quality, long term records exist, ideally located in the predictor site proximity, with similar climate; (3) obtain reference site wind data for the same time period as for the predictor site, the concurrent period; (4) determine the relationship between the reference and predictor site wind data for the concurrent period; (5) obtain wind data from the reference site for a historic period of over 10 years duration or the longest possible, the historic period; and (6) apply the relationship determined in step (4) to the historic data from the reference site to "predict" what the winds would have been at the predictor site over that period. These are the wind predictions that would have been observed and the measurements were made at the predictor site for the same period as the historic data, rather than a wind velocity prediction. The MCP key factor is the algorithm or relationship used in step (4). Most MCP techniques use direction sectored regression analysis to establish relationships of the wind speed and direction at the reference site and the ones at the potential wind farm site. The long-term wind data may be taken from nearby meteorological stations or data from the NCEP/NCAR reanalysis dataset. The general approach is to look for a relationship between the wind speed variables vsite and v_{met} of the site under development and a suitable reference station:

$$v_{site} = f(v_{met}) \tag{58}$$

Often, it may be suitable to consider several reference stations with concurrent data sets for a given development site; Eq. (57) can be then generalized to:

$$v_{site} = g\left(v_{met}^{1}, v_{met}^{2}, \dots, v_{met}^{N}\right) \tag{59}$$

Currently different MCP variants are implemented in the WindPRO, WAsP or other wind energy software packages. However, many researchers developed own MCP applications. Notice that the wind speed time series can be analyzed irrespective of wind direction, while usually, the wind direction is binned into a certain number of sectors and the wind speed subsets for each direction bin are analyzed separately for their MCP correlations. Since wind direction observations may not always coincide, binning may be based either on the wind direction measured at the reference station (usually) or at the prospect site. When a systematic veer occurs (e.g., in response to the topography), the relationship between the site and the meteorological station direction may be fitted, and the fit curve may be used to predict the long-term site wind direction. A deeper objective, requiring a more insightful analysis of the statistical behavior is to achieve an estimation of the probability density function $f_{Y,long}\ (v_{sim},\ v_{met},\ \alpha)$, which best describes the long-term wind speed distribution at a given location and height above ground level based on the knowledge of the density in the reference site. Here α is a vector of parameters for each one of the distributions; the number and type of parameters here depend on the particular distributions. The knowledge of $f_{Y,long}\ (v_{sim},\ v_{met},\ \alpha)$, allows to calculate the average wind speed and power density and, most importantly, the average energy yield of a given wind turbine [4, 72–77]. Additionally, intra- or inter-annual fluctuations of the wind resource at the prospective site need to be studied, although the present study does not consider such variations. The models employed in practical applications fall into two classes, linear and non-linear and can be described mathematically, adapting previous equations, as:

$$v_{pred} = f\left(v_{ref}, \theta_{ref}\right) \tag{60}$$

$$\theta_{pred} = g\left(v_{ref}, \theta_{ref}\right) \tag{61}$$

Here, f and g are the functional relationships between two concurrent data, target and reference sites. Subscripts denote the data set, the reference as ref., or predictor as pred. There are several variants of these models, each with advantages and disadvantages.

4.1 Linear and regression models

The regression MCP method holds the traditional linear regression MCP analysis as a specialized subset of regression models using polynomials of other orders. Polynomial fitting methods are included, as suggested in literature [73–77]. In the simplest linear models the wind speed and direction at the target/predictor site is expressed as:

$$v_{h,pred} = f\left(v_{h,ref}\right) \tag{62}$$

and

$$f\left(v_{h,ref}\right) = \frac{\overline{v}_{c,pred}}{\overline{v}_{c,pred}} v_{h,ref} \tag{63}$$

Here the over bar refers to average, while *c* and *h* to the concurrent and historic data. Regression, with one independent (x) and one dependent (Y) variable is expressed as:

$$Y = f(x) + e \tag{64}$$

Here, Y is the dependent variable, x is the independent variable, f(x) is the regression model, and e is a random error (residual). The regression model could be polynomials of any order or other models, but traditionally a linear model is assumed, as this model has been found to give reasonable fits for wind energy estimation. In the case of a regression MCP analysis, the independent variable could be the wind speed measured at the reference site. The dependent variable, Y is then the wind speed at the local WT site. The regression parameters are estimated through a least square algorithm. The distributions of the random errors may, reasonably be assumed to follow a zero mean Gaussian distribution, e ~ N(0,σ). However, the distribution of the residuals should be visually checked, so that the assumption is verified as reasonable. This is needed, as the random variable model for the residuals is included in the MCP model to give the right energy levels in the new MCP-corrected time series. Note, that currently the distribution of residuals is conditioned on the reference wind direction only. Thus, conditioned on the reference wind direction, the residuals are independent on the reference wind speed.

4.2 Matrix method

In the matrix methods the changes in the wind speed and direction (wind veer) are modelled through joint distributions fitted on the 'matrix' of wind speed bins and wind direction bins. The concurrent periods of the measured wind data are used to calculate the set of non-linear transfer functions, used for estimating wind speeds and directions from the reference site to the prospect site. Since real measurements suffer from data missing in bins in the dataset, this method needs a way to substitute the missing input bins. A basic assumption of the matrix method is that the long-term site data (wind speed and direction) can be expressed through the simultaneous measurements of on-site data and reference site data. How this joint distribution is modelled should actually depend on the data in question, suggesting that a combination of binned sample distributions and modelled joint Gaussian distributions are working well [73–77]. The transfer model, given as a conditional distribution, is actually the key distribution in the matrix method. When applying the matrix method this conditional distribution is stipulated to hold regardless of the time frame considered. Thus, for each measured sample it is necessary to calculate/measure pairs of the two quantities (a pair is data with identical timestamps):

$$\Delta v = v_{site} - v_{ref} \tag{65}$$

$$\Delta \theta = \theta_{site} - \theta_{ref} \tag{66}$$

These parameters refer to the wind speeds and directions at the wind project site and the meteorological site, respectively. The joint distribution of $f(\Delta u, \Delta \theta)$ is modelled conditioned on the wind speed and the wind direction on the reference site. The joint distributions are represented as either through the samples (bootstrap model) or often through a joint Gaussian distribution. When the data has been measured and a match between the short-term data and the short-term reference data has been established, then the samples are sorted into bins with specific resolutions, such as 1 m/s and 10 degrees. The result from this binning is a set of joint sample distributions of wind veer and wind speeds. Since the data is binned with wind speed and wind direction, these sample distributions are conditioned on the mean wind speed at the reference position and the wind direction on the reference position. The calculated distributions are used directly in a bootstrapping technique when doing matrix MCP calculations. Based on the sample distributions, the statistics, calculated for the wind veer are the mean, the standard deviation, and the correlation coefficients. To enable interpolations and extrapolations into bins where no data is present, a spline is fitted to the sample distributions. This parametric distribution is represented by the two moments and the correlation, assuming that a joint Gaussian distribution. Note, that even if the Gaussian distribution assumption may seem a bit crude, then the parametric model can be applied in cases where limited or no sample data is available. Thus, the influence of this assumption is limited, as most long-term corrected samples are typically based on the resampling approach. The mean, standard deviation and correlation are now modelled as 'slices' of polynomial surfaces:

$$P(v, \theta) = \sum_{i=0}^{N} a_i(v_i, \theta_i) v_{ref}^1 \tag{67}$$

where P denotes the sample statistical moment (or correlation) considered, N is the order of the polynomial is the polynomial coefficients, depending on the wind velocities. As in the case of regression MCP, the long-term corrected meteorological data is calculated using Bootstrap and Monte-Carlo techniques, i.e. probabilistic methods enabling generation of the long-term corrected wind distribution through an *artificial* time series.

4.3 Hybrid MCP and the wind index MCP methods

The hybrid MCP method [72–77] correlates the wind data at the targeted wind plant site with that at multiple reference stations. The strategy accounts for the local climate and the topography information. In the original hybrid MCP method, all component MCP estimations between the targeted wind plant site and each reference station use a single MCP method (e.g., linear regression, variance ratio, Weibull scale, or neural networks). The weight of each reference station in the hybrid strategy is determined based on: (i) the distance and (ii) the elevation differences between the target wind plant site and each reference station. The hypothesis here is that the weight of a reference station is larger when the reference station is closer (shorter distance and smaller elevation difference) to the target wind plant site. The weight of each reference station, w_i, is determined by:

$$w_j = F\left(n_{ref}, \Delta d_j, \Delta h_j\right) \tag{68}$$

where n_{ref} is the number of reference stations; and Δd_j and Δh_j represent the distance and the elevation difference between the target site and jth reference station, respectively. Each wind data point is allocated to a bin according to the wind direction sector at the target wind plant site. Within each sector, the long-term wind speed is predicted by a hybrid MCP strategy based on the concurrent short-term wind speed data within that specific sector. By setting the wind speed data in each sector together, the long-term wind data at the target wind plant site can be obtained. The predicted long-term wind data quality is usually evaluated using the performance metrics. ANNs are used to correlate and predict wind conditions because of their ability to recognize patterns in noisy or complex data. A neural network contains an input layer, one or more hidden layers, and an output layer, being defined the following parameters: input and output connections, number of neuron layers, the weights, and transfer functions, the interconnection pattern between different neuron layers, the learning process for updating the weights of the interconnections, and the activation function that converts the input into outputs. The Levenberg–Marquardt algorithm is usually used for neural network training.

The index correlation method is a method creating the MCP analysis by means of monthly averages of the energy yield, disregarding the wind directional distributions [72–77]. Even though this method may seem rather crude and primitive when comparing to other more advanced MCP methods, which takes the wind veer into account; this method has the advantages in stability and performance as it may even succeed in the cases where other MCP methods may fail. This is due to the fact, that the wind indexes are related directly to WTG energy yield and that the method allows the production calculation to be completed using actual measured data before applying the correction. The Wind Index MCP method offers the opportunity to calculate the wind indexes using real power curves of the wind turbines. A generic power curve based on a truncated squared wind speed approach may be chosen. When the wind indexes are calculated, the MCP correction is done on the estimated WTG energy yield, i.e. by multiplying the production estimated with a correction factor based on the difference in the wind index from the short-term site data to the long-term site estimated data. However, since the power curve of a WTG is a non-linear function of the wind speed the wind index is typically modelled using power curve common models. For the power output, calculated for the target site and the reference to be comparable they must be based on a similar mean wind speed. This is done by assuming a sector uniform shear that can be applied so that both concurrent mean wind speeds are set to a fixed user-inferred wind speed, typically the expected mean wind speed at hub height. The individual wind speed measurements are thus multiplied with the relevant factor. Full time series wind speeds are adjusted with the same ratio as the one applied to the respective concurrent time series. The argument for this operation is that the variations in wind speed will only be interpreted correctly in terms of wind energy if a comparable section of the power curve is considered.

As discussed in [4, 9, 14, 72–77], once a regression equation has been conditioned based on the measurement overlap period, the regression parameters can then be used to derive an extended data record for the site of interest. MCP methods are generally applied using some sort of regression analysis for each wind direction sector. An issue of using MCP methods based on wind velocity data from the land sites, due to the scarcity of offshore wind velocity observations is that most applications use linear regression which cannot account for observed differences in the wind speed distribution between the land site and the offshore sites. In [9, 14] was proposed in such cases to apply wind velocity corrections of the probability distribution functions, e.g. Weibull parameter corrections. In this method, the Weibull parameters of the short-term data series are modified to characterize longer data sampling periods. It compares sector-based wind speed distributions at the onshore and the off-shore sites considering the on-shore long-term time series as representative of the area of interests. Weibull scale (c) and shape (k) factors are determined for each of several wind sectors and for the mean values in each point of the grid, for data sets from overlapping periods. The differences between the two datasets are expressed in terms of the ratios of Eq. (69). These correction factors are applied to the Weibull factors estimated for the short-term data sets.

$$Corr_{Scale} = \frac{c(\text{off-shore})}{c(\text{on-shore})}, \text{ and } Corr_{Shape} = \frac{k(\text{off-shore})}{k(\text{on-shore})} \tag{69}$$

The Weibull correction method, as discussed in [9, 14] gave better wind velocity estimated for both onshore and offshore flow where the wind speeds were overestimated.

5. Wind energy resources in climate projections

Just as with the other aspects of climate, wind statistics are subject to natural variability on a wide range of time scales. Like other meteorological parameters, such as temperature, rainfall, or other climate variables, wind speeds and directions change on time scales of minutes, hours, months, years, and decades. Future climate change is expected to alter the spatio-temporal distribution of surface wind speeds and directions, with impacts on wind-based electricity generation. Long term trends in wind speeds are difficult to quantify and large historical data sets are required to accurately capture and describe such variations. This is a more evident in the case of the offshore wind energy resources. Wind energy resources at any location vary on a range of time scales, and hence any resource assessment should address issues of climate variability and change. However, due to scarcity of complete datasets offshore, comparisons have generally been performed with the hypothesis that local wind regimes have not changed during the last 10–20 years. The assumption validity is questionable, being likely to be regionally variable [78–81]. Even in the absence of climate non-stationarity, wind energy measurement sites typically have data periods up to 3 years and hence are not representative of wind climates over the 20–30 year lifetime of the wind farms. A further confounding influence is that homogeneous wind speed time series are rarely available for long periods because many monitoring locations have undergone change in land use and instrumentation. Accordingly, one can conceptualize the wind resource assessment as a two-step process: (1) an evaluation of wind resources at the regional scale to locate promising wind farm sites and (2) a site specific evaluation of wind climatology and vertical profiles of wind and atmospheric turbulence, in addition to an assessment of historical and possibly future changes due to climate variability.

In the context of wind energy generation, even small changes in the wind speed magnitude can have major impacts on the productivity of wind power plants, as the wind power relationship is directly proportional to the cube of the wind speed. However, the predictions for the direction and magnitude of these changes hinge critically on the assessment methods used. Decadal and multi-decadal variability in wind speed statistics currently introduce an element of risk into the decision process for siting new wind power generation facilities. Recent findings from the atmospheric science community suggest that climate change may introduce an added risk to this process. Many climate change impact analyses, including those focused on wind energy, use individual climate models and/or statistical downscaling methods rooted in historical observations. Wind speed and direction vary on small scales and respond in complex ways to changes in large-scale circulation, surface energy fluxes, and topography. Thus, whereas multiple climate models often agree qualitatively on temperature projections, wind estimates are less robust. The spatial variability of wind and its sensitivity to model structure suggest that higher resolution models and multi-model comparisons are particularly valuable for wind energy projections. For long-term planning of wind resources, it is imperative to analyze historical datasets and establish monitoring at hub-height using meteorological towers and remote sensing. A comprehensive review of climate change impacts on wind energy is shown in [78–81], discussing the main changes in the wind resources due to climate evolution, focused on northern Europe, with significant wind energy installations.

According to the analysis, until the middle of the current century natural variability will exceed the effect of climate change in the wind energy resources [78–81]. They conclude that there is no detectable trend in the wind resources that would impact future planning and development of wind industry in northern Europe. Pryor et al. (2006) down-scaled winds from ten global climate models at locations in northern Europe and found no evidence of significant changes in the 21st century wind regime compared to the 20th century. Predicted changes are found to be small and comparable to the variability associated with different global climate models. Using another approach, Ren [79] proposed a power-law relationship between global warming and the usable wind energy. The power-law exponent was calibrated using results from eight global climate models. He found that reduction of wind power scales with the degree of warming according to method and estimated that about 4 Celsius degrees increase in the temperatures in mid to high latitudes would result in up to 12% decrease in wind speeds in northern latitudes. Ren [79] suggested that an early maximized harvesting is beneficial and should be carried out. However, more studies are needed to solve all uncertainties in climate projections of wind resources under various future climate scenarios [4].

6. Chapter summary

Several factors are influencing the accurate assessment and prediction of wind energy production. A primary issue is adequate understanding of the effects of wind variability, atmospheric stability, turbulence and air density variability on the wind turbine energy production. Non-negligible and quite often significant error is incurred when the effects of shear, TI, and atmospheric stability on the wind turbine power performance are ignored, as in the IEC standard, 61400-12-1 (2005). The standard procedures are valid only for ideal neutral conditions and a small wind turbine. Besides the dominant cubic dependence of the wind speed on the wind power density, there are smaller but still important corrections to the air density that are important to harvesting wind energy at high-elevation sites. Corrections that account for these factors must be included in the power output estimates, and more accurate predictions will help alleviate production-consumption imbalances. These imbalances can also be ameliorated through the use of storage devices. The field of wind resource assessment is evolving rapidly, responding to the increasingly stringent requirements of large-scale wind farm projects often involving investments of several hundred million dollars. Traditional cup anemometry is being complemented with ultrasonic sensors providing information on all three components of the wind velocity vector and enabling a better assessment of turbulence. Remote sensing devices like sodar and lidar are becoming more popular as turbine hub heights and rotor diameters increase, often placing the upper edge of the swept rotor area at heights of 130 m or more. While the traditional, conventional approaches of measuring the wind speed and direction at a few heights below hub height and extrapolating based on a logarithmic profile is still very common, the use of both vertical profiling devices and more accurate modeling tools considering the full terrain complexity and atmospheric stability is quickly moving into the mainstream. Measure-correlate-predict (MCP) methods are used to estimate wind speeds and directions at a target site where wind power is assessed for development. These methods use two sets of in-data. To begin with a series of measured speeds and directions from the target site during a period of time (usually one year) is needed. In addition to this, a reference series from a much longer period needs to be obtained. On the other hand, the advanced hybrid MCP method uses the recorded data of multiple reference stations to estimate the long-term wind condition at a target plant site. Because each reference station has the flexibility to use any of the available MCP techniques, the multiple reference weather stations were combined into the hybrid MCP strategy with the best suitable MCP algorithm for each reference station. Climate projections of wind resources in changing climate are a topic of a debate in the literature, requiring a thorough investigation of uncertainties and understanding the complex interaction of atmospheric dynamics. This will contribute to understanding the extent to which some of the predicted trends are the result of the weather and climate variability or the result of inadequate physical parameterizations in global and regional climate models.

Author details

Radian Belu
Southern University and A&M College, Baton Rouge, USA

*Address all correspondence to: radian_belu@subr.edu

References

[1] Belu, R., Wind Energy Conversion and Analysis. In Anwar Sohail editor, Encyclopedia of Energy Engineering and Technology, Taylor and Francis, 2013, DOI: 10.1081/E-EEE-120048430, (27 pages).

[2] Belu R.G., Renewable Energy Systems: Fundamentals and Source Characteristics, Taylor and Francis, CRC Press, 420p, 2019, doi.org/10.1201/9780429297281

[3] Manwell, J. F,, McGowan, J. G., and Rogers, A. L., Wind energy explained : theory, design and application (2nd ed.),. Amherst, USA: John Wiley & Sons, 750p, 2009.

[4] Belu R.G and Koracin D., Wind Energy Analysis and Assessment, Advances in Energy Research, 2015:20 (1):1-55.

[5] Landberg, L., Myllerup, L., Rathmann, R. et al., Wind resource estimation-An overview. Wind Energy, 2003:6:261-271.

[6] Petersen, E. L., Mortensen, N. G., Landberg, L., Wind power meteorology. Part I: Climate and turbulence, Wind Energy, 1998: 1(1):25-45.

[7] Petersen, E. L., Mortensen, N. G., Landberg, L., Wind power meteorology, Part II: Siting and models, Wind Energy, 1998:1(2):55-72.

[8] Archer, C. L., Jacobson, M. Z., Spatial and temporal distributions of US winds and wind power at 80 m derived from measurements, J. Geophys. Res., 2003: 108:doi:10. 1029/2002/D002076. 84.

[9] Sempreviva, A. M., Barthelmie, R. J., and Pryor, S. C., Review of Methodologies for Offshore Wind Resource Assessment in European Seas, Surv Geophys, 2008:29:471–497 DOI 10.1007/s10712-008-9050-2

[10] Garcia-Bustamante, E., Gonzalez-Rouco, J. F., Jimenez, P. A., Navarro, J., and Montavez, J. P., The influence of the Weibull assumption in monthly wind energy estimation. Wind Energy, 2008: 11:483-502.

[11] Greene, J. S., Morrissey, M. L. and Johnson, S., Wind Climatology, Climate Change, and Wind Energy, Geography Compass, 2010: 4(11):1592-1605.

[12] Landberg, L., Myllerup, L., Rathmann, O., Petersen, E. L. et al., Review wind resource estimation - an overview, Wind Energy, 2003: 6: 261–271.

[13] Bechrakis, D., J. and McKeogh, D., Wind resource assessment of an area using short term data correlated to a long-term data set. Solar Energy, 2004: 76:725-732.

[14] Barthelmie, R., Hansen, O., Enevoldsen, K., Hoejstrup, J., Larsen, S. E., Frandsen, S., Pryor, S., Motta, M., Sanderhoff, P., Ten years of meteorological measurements for offshore wind farms. J. Sol. Energy Eng., 2005: 127(2):170–176. doi:10.1115/1.1850489

[15] Jiang, Q., Doyle, J. D. Haack, T., Dvorak, M. J., Archer, C. L. and Jacobson, M. Z., Exploring wind energy potential off the California coast, Geophys. Res. Lett., 2008:35: L20819: doi:10.1029/2008GL034674.

[16] Jimenez, B., Durante, F., Lange, B., Kreutzer, T., and Tambke, J., Offshore wind resource assessment with WAsP and MM5: Comparative study for the German Bight, Wind Energy, 2007:10: 121-134.

[17] Dhanju A., Whitaker, P., and Kempton, W., Assessing offshore wind resources: An accessible methodology. Renewable Energy 2008:35: 1244–1254.

http://www.sciencedirect.com/science/article/pii/S096014810700078X.

[18] Gasset, N., Landry, M., Gagnon, Y., A Comparison of Wind Flow Models for Wind Resource Assessment in Wind Energy Applications, Energies, 2012:5: 4288-4322: doi:10.3390/en5114288.

[19] IEC Standard, 6-1400-12-1, 2005: Power performance measurements of electricity producing wind turbines, 158.132.178.190 › IEC61400part12_1_WindMeasurement [Accessed: 09-20-2020].

[20] Antoniou, I., Pedersen, S., Enevoldsen, P.D., Wind shear and uncertainties in power curve measurement and wind resources, Wind Engineering, 2009:33: 449-468.

[21] Fırtın, E., Guler, O., and Akdag, S. A., Investigation of wind shear coefficients and their effect on electrical energy generation, Applied Energy, 2011:88:4097-4105.

[22] Kwon, S., Uncertainty analysis of wind energy potential assessment, Applied Energy, 2010:87(3):856-865.

[23] Justus, C. G., Mikhail, A, Height variations of wind speed and wind distribution statistics, Geophys. Res. Lett., 1976:3:261-264.

[24] Peterson, E. W., J. P. Hennessey, On the use of power laws for estimates of wind power potential, J. Appl. Meteorol., 1977: 17:390-394.

[25] Weggel, R. J. Maximum daily wind gust related to mean daily wind speed. J. Struct. Eng., 1999:125: 465-469.

[26] Sisterson, D. L., B. B. Hick, R. L. Coulter, M. L. Wesely, Difficulties in using power laws for wind energy assessment, Solar Energy, 1983: 31, 201-204.

[27] Smedman, A., U. Hogstrom, H. Bergstrom, Low level jets - a decisive factor for offshore wind energy siting in the Baltic Sea, Wind Engineering,1996: 28:137-147.

[28] Sumner, J., Masson, C., Influence of atmospheric stability on wind turbine power performance curves, J. Sol. Energy Eng., 2006:128:531-537.

[29] Wagner, R., I. Antoniou, S. M. Pedersen, M. S. Courtney, H. E. Jørgensen, The influence of the wind speed profile on wind turbine performance measurements, Wind Energy, 2009:12;348-362.

[30] Wharton, S., J. K. Lundquist, Atmospheric stability affects wind turbine power collection, Environ. Res. Lett., 2011:7: 014005 (9 pp.).

[31] US DOE, 2008: Research needs for wind resource characterization. NREL Tech. Rep. 500-43521, 116 pp. [Available online at www.nrel.gov/docs/fy08osti/43521.pdf].

[32] Vanderwende, B. J. and Lundquist, J. K., The modification of wind turbine performance by statistically distinct atmospheric regimes. Environ. Res. Lett., 2011:7:03403 5.

[33] Barthelmie, R. J., S. C. Pryor, S. T. Frandsen, K. S. Hansen, J. G. Schepers, K. Rados, W. Schlez, A. Neubert, L. E. Jensen, and S. Neckelmann, Quantifying the impact of wind turbine wakes on power output at offshore wind farms. J. Atmos. Oceanic Technol., 2010: 27"1302-1317, doi:10.1175/2010JTECHA1398.1.

[34] Belu, R. G., and Koracin, D., Effects of the Complex Wind Regimes and Meteorological Parameters on Wind Turbine Performances, IEEE Energy Tech. 2012 Conference, May 29-31, 2012, Cleveland, Ohio (CD Proc.).

[35] Probst, O., Cárdenas, D., State of the Art and Trends in Wind Resource

Assessment, Energies, 2010: 3: 1087-1141; doi:10.3390/en3061087.

[36] Lang, S., McKeogh, M., LIDAR and SODAR Measurements of Wind Speed and Direction in Upland Terrain for Wind Energy Purposes, Remote Sens. 2011:3:1871-1901; doi:10.3390/rs3091871.

[37] Peña, A., C. B. Hasager, S. Gryning, M. Courtney, I. Antoniou, and T. Mikkelsen, Offshore wind profiling using light detection and ranging instruments. Wind Energy, 2009:12: 105-124. doi: 10.1002/we.283.

[38] Pichugina, Y. L., Robert M., et al., Doppler Lidar–Based Wind-Profile Measurement System for Offshore Wind-Energy and Other Marine Boundary Layer Applications. J. Appl. Meteor. Climatology, 2012: 51:327-349. doi: doi: 10.1175/JAMC-D-11-040.1.

[39] Hennessy, J. P., 1977: Some aspects of wind power statistics, J. of Appl. Meteor., 1977:16:119-128.

[40] Justus, C. G., Hargraves, W. R., Mikhail, A., Graber, D., Methods of estimating wind speed frequency distribution, J. of Appl. Meteorology, 1978: 17: 350-353.

[41] Crotis, R. B., Sigl, A. B., Klein, J., Probability models of wind velocity magnitude and persistence, Solar Energy, 1978:20: 483-93.

[42] Kaminsky, F. C., Four probability densities (log-normal, gamma, Weibull, and Rayleigh) and their application to modelling average hourly wind speed. In: Proceedings of the International Solar Energy Society, 1977: 19.6-19.10.

[43] Auwera, L., Meyer, F., Malet, L., The use of the Weibull three-parameter model for estimating mean wind power densities, J. Applied Meteorology, 1980: 19:819-825.

[44] Deaves, D. M., Lines, I. G., On the fitting of low mean wind speed data to the Weibull distribution, J. Wind Eng. and Industrial Aerodynamics, 1997: 65: 169-178.

[45] Ulgen, K., Genc, A., Hepbasli, A., et al., Assessment of wind characteristics for energy generation, Energy Sources, 2004:26(13):1227-1237.

[46] Monahan, A. H., The probability of sea surface wind speeds, part I: Theory and Sea Winds Observations, J. Climate, 2006:19:497-520.

[47] Basumatary, H., Sreevalsan, E., and Sai, K. K., Weibull parameter estimates – a comparison of different methods, Wind Engeering, 2005:29:309-315.

[48] Hirose, H., Maximum likelihood parameter estimation in the three-parameter gamma distribution, Comput. Stat. Data Anal., 1995:20(4): 343-54.

[49] Pryor, S. C., and Barthelmie, R. J., Statistical analysis of flow characteristics in the coastal zone. J. Wind Engineering and Industry Aerodynamics,2002: 90(3):201-221.

[50] Ramirez, P., and Carta, J. A., Influence of the data sampling interval in the estimation of the parameters of the Weibull wind speed probability density distribution: a case study. Energy Conversion and Management, 2005: 46(15-16):2419-2438.

[51] Belu, R. G., and Koracin, D., Wind Characteristics and Wind Energy Potential in Western Nevada, Renewable Energy, 2009:34(10): 2246-2251.

[52] Belu, R. G., and Koracin, D., Statistical and Spectral Analysis of the Wind Characteristics in the Western Nevada, J. of Wind Energy, 2013:(12 pages), Article ID 739162, 2013. doi: 10.1155/2013/739162.

[53] Carta, J. A., and Ramirez, P., Analysis of two-component mixture Weibull statistics 0for estimation of wind speed distributions, Renew. Energy, 2007: 32(3): 518-531.

[54] Chang, T. P., Performance comparison of six numerical methods in estimating Weibull parameters for wind energy application, Applied Energy, 2011: 88:272-282.

[55] Seguro, J. V., and Lambert, T. W., Modern estimation of the parameters of the Weibull wind speed distribution for wind energy analysis, J. Wind Eng. and Industry Aerodynamics, 2000:85:75- 85.

[56] Beccali, M., Cirrincione, G., Marvuglia, A., and Serporta, C., Estimation of wind velocity over a complex terrain using the Generalized Mapping Regressor, Applied Energy, 2010:87: 884-893.

[57] Akdag, S. A., and Dinler, A., A new method to estimate Weibull parameters for wind energy applications. Energy Convers. Management, 2009: 50: 1761-1766.

[58] Akdag, S. A., Bagiorgas, H. S., and Mihalakakou, G., Use of two-component Weibull mixtures in the analysis of wind speed in the Eastern Mediterranean, Applied Energy, 2010: 87:2566-2573.

[59] Chang, T. P., 2011; Estimation of wind energy potential using different probability density functions, Applied Energy, 2011:88:848-1856.

[60] Rocha, P. A. C., Coelho de Sousa, R., Freitas de Andrade, C., et al., Comparison of seven numerical methods for determining Weibull parameters for wind energy generation in the northeast region of Brazil, Applied Energy, 2012:89:395-400.

[61] Carta, J. A., and Velázquez, S., A new probabilistic method to estimate the long-term wind speed characteristics at a potential wind energy conversion site, Energy, 2011: 36: 2671-2685.

[62] Villanueva, D., and Feijoo, A., Wind power distributions: A review of their applications, Renewable and Sustainable Energy Reviews, 2010:14:1490-1495.

[63] Morgan, E. C., Lackner, M., Vogel, R., and Baise, L. G., Probability distributions for offshore wind speeds, Energy Conversion and Management, 2011:52:15-26.

[64] Li, M. and Li, X., MEP-Type Distribution Function: A Better Alternative to Weibull Function for Wind Speed Distributions, Renewable Energy, 2005:30(8): 1221-1240. doi: 10.1016/j.renene.2004.10.003

[65] Akpinar, S., Akpinar, E. K., Estimation of wind energy potential using finite mixture distribution models, Energy Convers. Manage. 2009: 50(4):877-884.

[66] Carta, J. A., and Ramirez, P., Use of finite mixture distribution models in the analysis of wind energy in the Canarian Archipelago. Energy Conversion and Management, 2007:48(1):281-291.

[67] Zhang, J., Chowdhury, S., Messac, A., and Castillo, L., A multivariate and multi-modal wind distribution model, Renewable Energy, 2013:51:436-447.

[68] Ramirez, P., Carta, J. A., 2006: The use of wind probability distributions derived from the maximum entropy principle in the analysis of wind energy: a case study, Energy Convers. Management, 2006:47(15-16):2564-2577.

[69] Vicente, R-T., Influence of the fitted probability distribution type on the annual mean power generated by wind turbines: A case study at the Canary Islands, Energy Convers. Management, 2008:49:2047-2055.

[70] Carta, J. A., Ramírez, P., and Velázquez, S., Influence of the level of fit of a density probability function to wind-speed data on the WECS mean power output estimation, Energy Convers. Management, 2008:49(4): 2647–2655

[71] Kantar, Y.M., and Usta, I., Analysis of wind speed distributions: wind distribution function derived from minimum cross entropy principles as better alternative to Weibull function, Energy Convers. Management, 2008:49: 2647-2655.

[72] Nawri, N., Petersen, G. N., Bjornsson, H. et al., The wind energy potential of Iceland, Renewable energy, 2014:69:290-299.

[73] Zhang, J., Chowdhury, S., Messac, A., and Castillo, L., A hybrid measure-correlate-predict method for wind resource assessment, In: ASME 2012 6th International Conference on Energy Sustainability, 2012 ASME Conference Proceedings.

[74] Andersen, M., A review of measure-correlate-predict techniques, Technical Report No. 01327R00022, Renewable Energy Systems, 2004.

[75] Rogers, A. L., Rogers, J. W., James, F., and Manwell, J. F., Comparison of the performance of four measure–correlate–predict algorithms. J. Wind Eng. Ind. Aerodyn., 2005:93:243-264.

[76] Perea, A. R., Amezcua, J. and Probst, O. 2011: Validation of three new measure-correlate-predict models for the long-term prospection of the wind resource. Journal of Renewable and Sustainable Energy, 2011:3(2):023105.

[77] Lackner, M. A., Rogers, A. L. and Manwell, J. F., 2008: Uncertainty analysis in mcp-based wind resource assessment and energy production estimation, Journal of Solar Energy Engineering, 2008:130(3): 031006.

[78] Pryor SC, Barthelmie RJ, Kjellstro¨m E., Analyses of the potential climate change impact on wind energy resources in northern Europe using output from a Regional Climate Model. Clim Dyn., 2005:25:815–835. doi: 10.1007/s00382-005-0072-x

[79] Ren, D., Effects of global warming on wind energy availability. J. Renewable Sustainable Energy 2010:2, 052301; doi: 10.1063/1.3486072

[80] Klink, K., Climatological mean and interannual variance of United States surface wind speed, direction and velocity, Int. J. Climatol., 1999: 19: 471-488.

[81] Koracin D., R.G. Belu. K. Horvath, B. Candillas, J. Jainag, and G. Smith, A review of challenges in assessment and forecasting of wind energy resources, Croatian Meteorological Journal, 2012: 47:13-33.

Chapter 9

Nonlinear Control Strategies of an Autonomous Double Fed Induction Generator Based Wind Energy Conversion Systems

Nouha Bouchiba, Souhir Sallem, Mohamed Ben Ali Kammoun, Larbi Chrifi-Alaoui and Saïd Drid

Abstract

In the last few decades, among the wide range of renewable energy sources, wind energy is widely used. Variable speed wind energy conversion systems based on double fed induction generator have a considerable interest mostly in case of islanded networks and/or isolated applications. In this paper, as a means to supply remote areas, an investigation of a wind energy conversion system (WECS) based on a double fed induction generator (DFIG) is carried out. The presence of both wind turbine aerodynamics and DFIG coupled dynamics causes strong nonlinearities in the studied system. Wind speed and demanded power variations have a major impact on the quality of the produced energy. In order to control and maintain the stator output voltage and frequency at their nominal values (220 V/50 Hz) under wind speed and load variations, this work presents a study of three kinds of controllers: PI, Back-Stepping and Sliding Mode controllers. These controllers are integrated in the studied system and a comparison of their dynamic performances has been developed. Moreover, in order to ensure the rotor side converter safety on the one hand and to guarantee an optimal operation of the DFIG on the other hand, a management strategy is proposed in this work. Simulation results are performed using Matlab/Simulink environment and show the effectiveness and the accuracy of each controller compared to others mainly with the presence of wind speed and load demand variations.

Keywords: wind energy conversion system, double fed induction generator, PI controller, back-stepping controller, sliding mode controller

1. Introduction

Over the past two decades, one of the most important aspects of our life is electrical energy [1, 2]. Currently, to supply power for a modern life as well as to avoid environmental issues originating from fossil fuels exploitations, the production of clean energy has become the primary objective of major universal power producing nations [3, 4]. Solar energy, wind power, biomass and geothermal are the most useful renewable energy sources [5–7]. Nowadays, wind power has become a crucial renewable energy source [8–10].

To convert wind energy into electric power, many kinds of generator concepts have been used [11, 12]. Previously, the squirrel cage induction generator was basically used in wind energy conversion system [13]. This technology is well known in fixed speed applications. Recently, the technology moves towards variable speed wind energy conversion systems [14].

Thanks to its advantages such as four quadrant power capabilities, variable speed operation, improved efficiency, decoupled regulation and reduced losses, the Doubly Fed Induction Generator DFIG has been extensively used in wind energy conversion systems WECS [1, 12, 15, 16].

In fact, using this concept, the electronic power converters are designed only at 25 to 30% of the generator capacity [12]. Therefore, from an economic point of view, this technology is more attractive compared to others (PMSG) [9, 11, 17]. Moreover, a study demonstrates that the DFIG topology presents 50% of the wind power market [10, 15].

Instead of DFIG based WECS in grid connected operation mode, a very little consideration has been paid towards the stand-alone strategy where consumers are totally disconnected from the distribution network [12]. However, in the last few decades, the availability of electricity problem in remote areas has created opportunities to exploit renewable energy sources to feed isolated loads [15]. Therefore, the implementation of stand-alone or isolated power systems can handle the rural electrification for small or medium power consumers located far from the distribution grid by providing sustainable and reliable energy supply [10, 15, 18].

Owing to the large extension of the doubly fed induction generator in isolated power systems as a primary power source generator for handling the electrification requirements of numerous isolated consumers worldwide, stand-alone wind power systems based on DFIG have become one of the most promising used technologies [10, 15]. From this perspective, the modeling and the control of WECS based DFIG have attracted extensive research efforts [2, 10, 15, 16].

Under variations of wind speed and power demand, the stator output voltage and frequency are no longer constant [12]. Highly fluctuating and unpredictable wind generation can have consequences in terms of system stability and robustness [19]. In fact, the DFIG based on WECS has strong nonlinearities [20] and the stability of power system is confronted to new challenges [21]. Many kinds of control strategies are studied and developed in literature [3, 5, 8, 10, 15].

In [8], the vector control strategy is used for the purpose of control both the active and reactive powers. In fact, to ensure the optimal operation mode, a strategy based on Adaptive fuzzy gain scheduling of the PI controller is developed.

In [10], the application of the DFIG for an isolated wind power system is examined to supply the remote area using Double-fed Induction Generator. The objective of this study is to supply different loads such as balanced, unbalanced and nonlinear loads. This study investigates the application stator/load side converter for load harmonics mitigation in the studied system. The shunt active power filter function is added to the convention control scheme of the load/stator-front voltage source converter so as to improve load harmonics. A simple technique for rotor side converter is invested to regulate Voltage and Frequency at stator/load terminals.

In [15], authors developed a speed-sensor less control strategy for a stand-alone doubly fed induction generator supplying energy to an isolated load. This technique is based on the root mean square (rms) detection. This developed direct voltage control method is applicable for not only the balanced and unbalanced load but also for standalone and grid connected mode. The control of load side converter is beyond the scope of this paper and only a diode rectifier is used for the purpose.

In [18], using fuzzy PI controller, the authors have described a control strategy for variable speed wind turbine based on DFIG. The main goal of this work is to analyze, apply and compare two kinds of controllers such as classical and Fuzzy PI.

In [19], a technique of terminal voltage build-up and the control of a stand-alone WECS based on DFIG is described. This technique is based mainly on the pitch control of the wind turbine. The active and reactive output powers are controlled and maintained equal to their reference values under sudden perturbations of wind speed and/or load variations.

In this chapter, an improved structure of a variable speed stand-alone WECS based on DFIG is proposed. In this context, a general model of the wind turbine is displayed. A detailed analysis of the autonomous DFIG for transient stability analysis is performed. The main goal of the present work is to control the stator outputs voltage and frequency as well as to maintain them within permissible operational limits (220 V/50 Hz) under wind speed and load demand variations on the one hand and to ensure the rotor side converter security on the other hand. Accordingly, to achieve these purposes, three types of controllers have been explored, modeled and integrated into the global system: The classical PI controller, a Back-Stepping and a Sliding mode controller. Besides, a management strategy is suggested to guarantee the rotor side power under 30% of the DFIG nominal power. The implementation of the overall system and different controller designs with the management strategy is achieved using Matlab/Simulink environment. In fact, simulation performances analysis of the stand-alone DFIG based WECS using the classical PI controller, back-stepping and sliding mode controllers are exhibited and discussed. A comparison between different controller process performances under sudden variation of load and wind speed disturbances is presented.

The remaining parts of this paper are organized as follows. Section 2 depicts different system component models. Section 3 describes various controller designs. Section 4 highlights the controllers management strategy. Section 5 demonstrates and compares simulation results of used controllers. Finally, Section 6 presents some drawn conclusions.

2. System modeling

The simplified schematic of the studied system is shown in **Figure 1**. It consists of a stand-alone double fed induction generator driven by a variable speed wind turbine through a gearbox. The stator of the machine is directly connected to an isolated three phase resistive inductive load (R_{Lp}, L_{Lp}). In fact, in order to make the stator outputs voltage and frequency independent from the load demand changes as well as the rotational speed variations, the rotor of the machine is supplied through a rotor side controller followed by a rotor side converter. The modeling of different components is presented and explained subsequently.

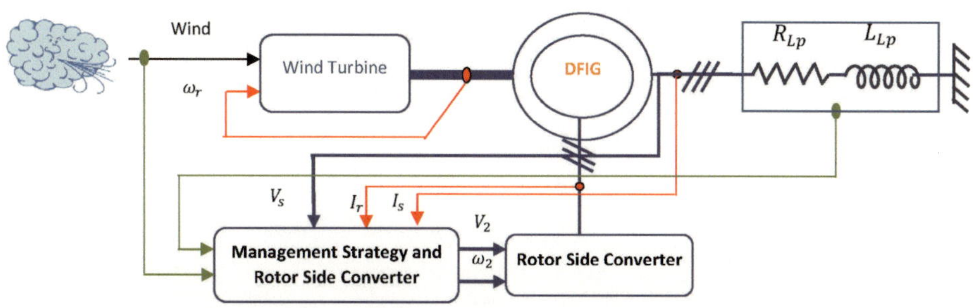

Figure 1.
Block diagram of the autonomous WECS based on DFIG.

2.1 Wind turbine model

The wind turbine aerodynamic modeling can be determined based on the power speed characteristics [12]. The mathematical expression of the mechanical power from wind turbine to the aerodynamic rotor is set forward (Eq. (1)).

$$P_m = \frac{1}{2}\rho\, R\, C_p\,(\lambda,\beta) V^3 \tag{1}$$

The power coefficient can be expressed in terms of the tip speed ratio and the pitch angle as follows (Eq. (2)):

$$C_p(\lambda, \beta) = C_1\left(\frac{C_2}{\lambda_i} - C_3\beta - C_4\right)e^{\frac{-C_5}{\lambda_i}} + C_6\lambda_i \tag{2}$$

Where the power coefficients are C_1 = 0.5176, C_2 = 116, C_3 = 0.4, C_4 = 5, C_5 = 21, C_6 = 0.0068 [12].

λ_i can be expressed by the following equation (Eq. (3)):

$$\frac{1}{\lambda_i} = \frac{1}{\lambda + 0.08\beta} - \frac{0.035}{\beta^3 + 1} \tag{3}$$

The tip speed ratio is given by (Eq. (4)):

$$\lambda = \frac{R\Omega_t}{V} \tag{4}$$

The aerodynamic torque, the generator torque and mechanical speed appearing on the shaft of the generator [11, 12] are represented respectively by (Eq. (5)–(7)).

$$C_{aer} = \frac{P_m}{\Omega_t} \tag{5}$$

$$C_m = \frac{C_{aer}}{G} \tag{6}$$

$$\Omega_t = \frac{\Omega_{mec}}{G} \tag{7}$$

In fact, the power of the used generator is low, the use of the pitch control can increase the cost of the whole system. Therefore, in this work, the pitch control does

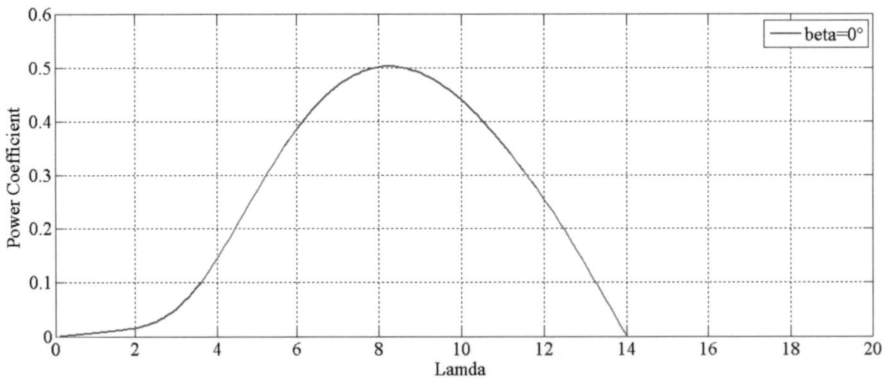

Figure 2.
Power coefficient for a pitch angle $\beta = 0$.

not prove to be a relevant solution to achieve our purpose and consequently, the pitch angle is maintained fixed ($\beta = 0$), which is a valid hypothesis for low and medium wind speeds [9]. **Figure 2** illustrates the power coefficient C_p variation versus the tip speed ratio λ for a specific chosen value of the pitch angle $\beta = 0$.

2.2 Standalone DFIG model

In this section, we attempt to analyze properly the DFIG in autonomous mode. The modeling of the three phase DFIG is carried out in a (d, q) reference frame. Using the generator convention, the Park model of the DFIG describing the functioning of this machine [22] in both stator and rotor side is given below respectively (Eqs. (8) and (9)):

$$\begin{cases} \frac{d}{dt}\psi_{sd} = -R_s I_{sd} - V_{sd} + \omega_1 \psi_{sq} \\ \frac{d}{dt}\psi_{sq} = -R_s I_{sq} - V_{sq} - \omega_1 \psi_{sd} \end{cases} \tag{8}$$

$$\begin{cases} \frac{d}{dt}\psi_{rd} = -R_r I_{rd} + V_{rd} + \omega_2 \psi_{rq} \\ \frac{d}{dt}\psi_{rq} = -R_r I_{rq} + V_{rq} - \omega_2 \psi_{rd} \end{cases} \tag{9}$$

As the d and q axis are magnetically decoupled, stator and rotor machine flux are expressed as (Eqs. (10) and (11)):

$$\begin{cases} \psi_{sd} = L_s I_{sd} + M I_{rd} \\ \psi_{sq} = L_s I_{sq} + M I_{rq} \end{cases} \tag{10}$$

$$\begin{cases} \psi_{rd} = L_r I_{rd} + M I_{sd} \\ \psi_{rq} = L_r I_{rq} + M I_{sq} \end{cases} \tag{11}$$

While functioning as a generator, the electromagnetic torque produced by the DFIG can be represented in terms of stator and rotor currents and flux as follows (Eq. (12)):

$$C_{em} = \frac{3}{2} p \frac{M}{L_s} \left(I_{rd}\psi_{sq} - I_{rq}\psi_{sd} \right) \tag{12}$$

Neglecting the machine viscous friction phenomenon, the electromechanical equation is given by (Eq. (13)):

$$\frac{dw_r}{dt} = \frac{p}{J}(C_m - C_{em}) \tag{13}$$

2.3 Load model

In stand-alone technology of DFIG based WECS, the stator of the machine is not connected to the grid but supplies an isolated load. Different kinds and values of loads can be connected to the stator terminals. The connected load is detected (R_{LP}, L_{LP}) in the following work. This couple (R_{LP}, L_{LP}) depends mainly on load demand power percentage noted LP_d [12]. Based on LP_d, the connecting load can be computed using the following equations (Eq. (14) and (15)).

$$R_{LP} = \frac{3V_{1n}^2}{LP_d.P_{1n}\sqrt{1+(tg\varphi^2)}} \tag{14}$$

$$L_{LP} = \frac{R_{LP}tg\varphi}{\omega_{1n}} \tag{15}$$

Where P_{1n}, V_{1n}, and w_{1n} represent respectively nominal power, voltage and pulsation of the machine.

Electrical equations on the stator side can be rewritten as follows (Eq. (16)) [12]:

$$\begin{cases} V_{sd} = R_{LP}i_{sd} + L_{LP}\frac{d}{dt}i_{sd} - \omega_1 L_{LP}i_{sq} \\ V_{sq} = R_{LP}i_{sq} + L_{LP}\frac{d}{dt}i_{sq} + \omega_1 L_{LP}i_{sd} \end{cases} \tag{16}$$

2.4 Converter model

For the considered power schema shown in **Figure 1**, the voltages across a, b, c rotor weddings of the DFIG are constructed as follows (Eq. (17)) [17].

$$\begin{bmatrix} V_{an} \\ V_{bn} \\ V_{cn} \end{bmatrix} = \frac{V_{dc}}{3}\begin{bmatrix} 2 & -1 & -1 \\ -1 & 2 & -1 \\ -1 & -1 & 2 \end{bmatrix}\begin{bmatrix} f_1 \\ f_2 \\ f_3 \end{bmatrix} \tag{17}$$

Where f_1, f_2 *and* f_3 represent the control signals and V_{dc} is the DC-link voltage referring to [17].

3. Controllers synthesis

The target of the proposed DFIG based WECS control is to keep the stator voltage amplitude and frequency constant and equal to their nominal values namely 220 V and 50Hz versus the load variations and wind speed fluctuations. Accordingly, the integration of a controller inside the studied system seems crucial. A few technologies about voltage and frequency control in an autonomous system based on DFIG are studied in literature [15]. However, in terms of complexity, these techniques exhibit many disadvantages in practice. In this work, we attempt to explore three types of controllers. In a first step, the implementation of the classical PI control technique is presented. Then, thanks to its advantages, the Back-Stepping controller is modeled and integrated into the system. Finally, a new technique of control is studied known as Sliding Mode controller.

In addition, the synthesis of different control strategies is based on choosing a synchronously dq reference frame with the stator voltage that is oriented with the d axis [23]. Consequently, we can formulate (Eq. (18)):

$$\begin{cases} \mathbf{V_{sd} = 0} \\ \mathbf{V_{sq} = V_s} \end{cases} \tag{18}$$

3.1 PI controller parameters calculation

In order to ensure the convergence conditions of the proposed system and to obtain good responses, PI controller parameters should be chosen properly. This

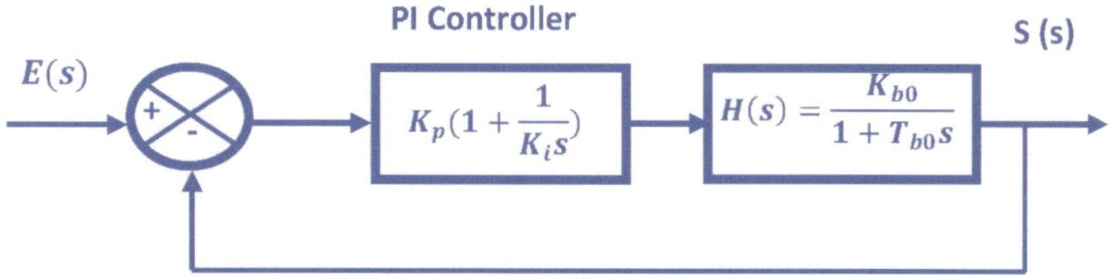

Figure 3.
The control diagram.

section describes a simple method for PI parameters computing. In general, the control diagram is presented as shown in **Figure 3**.

Where $H(s)$ represents the transfer function of the system which is given by (Eq. (19)):

$$H(s)=\frac{K_{b0}}{1+T_{b0}s} \tag{19}$$

K_p and K_i are the PI controller parameters for proportional and integral actions respectively. These parameters are computed based on DFIG parameters so as to ensure quick and convergent response of the DFIG based WECS subsystems. K_i is determined using the pole compensation method and K_p is deduced so as to ensure a fast response of the DFIG subsystems.

3.2 PI controller design

The main principle of the PI control topology is to control and regulate different physical parameters of the system using closed loops control. In this context, while applying this controller, in order to obtain the final control signals (U_{rd} and U_{rq}), two control loops are required. The computing of reference rotor currents (I_{rd} and I_{rq}) is carried out in a first step based on the calculation of the difference between reference and measured value of stator voltage (U_{sd} and U_{sq}). Calculating the output rotor voltages is performed in a second step by minimizing the error between reference and measured rotor currents already calculated in the first step. Therefore, stator and rotor voltages can be reformulated as follows (Eq. (20) and (21)):

$$\begin{cases} V_{sd}=\left[R_s I_{sd}+L_s\frac{dI_{sd}}{dt}-\omega_s L_s I_{sq}\right]+M\frac{dI_{rd}}{dt}-\omega_1 M I_{rq} \\ V_{sq}=\left[R_s I_{sq}+L_s\frac{dI_{sq}}{dt}+\omega_s L_s I_{sd}\right]+M\frac{dI_{rq}}{dt}+\omega_1 M I_{rd} \end{cases} \tag{20}$$

$$\begin{cases} V_{rd}=R_r I_{rd}+L_r\frac{dI_{rd}}{dt}+M\frac{dI_{sd}}{dt}-\omega_2 L_r I_{rq}-\omega_2 M I_{sq} \\ V_{rq}=R_r I_{rq}+L_r\frac{dI_{rq}}{dt}+M\frac{dI_{sq}}{dt}+\omega_2 L_r I_{rd}+\omega_2 M I_{sd} \end{cases} \tag{21}$$

The orientation of stator voltages with the d axis leads to (Eq. (22)):

$$\begin{cases} 0=R_s I_{sd}+L_s\frac{dI_{sd}}{dt}+M\frac{dI_{rd}}{dt}-\omega_1 L_s I_{sq}-\omega_1 M I_{rq} \\ V_s=R_s I_{sq}+L_s\frac{dI_{sq}}{dt}+M\frac{dI_{rq}}{dt}+\omega_1 L_s I_{sd}+\omega_1 M I_{rd} \end{cases} \tag{22}$$

Departing from these equations and omitting coupling terms, reference rotor currents can be expressed in terms of stator voltages.

Moreover, referring to (Eq. (22)), the first derivative of the stator current can be written as (Eq. (23)):

$$\begin{cases} \dfrac{dI_{sd}}{dt} = \dfrac{1}{L_s}\left[-R_sI_{sd} - M\dfrac{dI_{rd}}{dt} + \omega_1 L_s I_{sq} + \omega_1 M I_{rq}\right] \\ \dfrac{dI_{sq}}{dt} = \dfrac{1}{L_s}\left[V_s - R_sI_{sq} - M\dfrac{dI_{rq}}{dt} - \omega_1 L_s I_{sd} - \omega_1 M I_{rd}\right] \end{cases} \tag{23}$$

Referring to (Eq. (21)) and (Eq. (23)), reference rotor voltages are given as (Eq. (24)):

$$\begin{cases} V_{rd} = R_rI_{rd} - \omega_2 L_r I_{rq} - \omega_2 M I_{sq} + L_r\dfrac{dI_{rd}}{dt} + \dfrac{M}{L_s}L_d \\ V_{rq} = R_rI_{rq} + \omega_2 L_r I_{rd} + \omega_2 M I_{sd} + L_r\dfrac{dI_{rq}}{dt} + \dfrac{M}{L_s}L_q \end{cases} \tag{24}$$

Where

$$\begin{cases} L_d = \left[-R_sI_{sd} - M\dfrac{dI_{rd}}{dt} + \omega_1 L_s I_{sq} + \omega_1 M I_{rq}\right] \\ L_q = \left[V_s - R_sI_{sq} - M\dfrac{dI_{rq}}{dt} - \omega_1 L_s I_{sd} - \omega_1 M I_{rd}\right] \end{cases}$$

(Eq. (24)) can be rewritten as follows:

$$\begin{cases} V_{rd} = \left[R_rI_{rd} + \left(L_r - \dfrac{M^2}{L_s}\right)\dfrac{dI_{rd}}{dt}\right] + T_d \\ V_{rq} = \left[R_rI_{rq} + \left(L_r - \dfrac{M^2}{L_s}\right)\dfrac{dI_{rq}}{dt}\right] + T_q \end{cases} \tag{25}$$

Where

$$\begin{cases} T_d = -\left(\omega_2 L_r - \dfrac{M^2}{L_s}\omega_1\right)I_{rq} - M(\omega_2 - \omega_1)I_{sq} - \dfrac{R_sM}{L_s}I_{sd} \\ T_q = \left(\omega_2 L_r - \dfrac{M^2}{L_s}\omega_1\right)I_{rd} + M(\omega_2 - \omega_1)I_{sd} - \dfrac{R_sM}{L_s}I_{sq} + \dfrac{M}{L_s}V_s \end{cases}$$

3.3 Backstepping controller design

In this paper, we aim at improving performances of the studied system. In this context, in order to answer this need and to respond to our objective, we are basically interested in developing control strategies resting on linearization of the autonomous WECS based DFIG.

With the presence of many kinds of uncertainties, the back-stepping controller is able to linearize effectively a nonlinear system. In fact, during stabilization, unlike other techniques of linearization, this control technique has the flexibility to keep useful nonlinearities [23].

The stabilization of the virtual control state stands for the main purpose of the Backstepping controller. Therefore, this control strategy rests on the stabilization of a variable error by selecting carefully the suitable control inputs which are obtained from the analysis of Lyapunov function [24, 25].

In order to regulate effectively the output reference rotor voltages, the reference rotor currents are obtained based on a PI controller. Then, the rotor voltages are obtained by using a back-stepping controller.

Indeed, the first step of the Backstepping control is meant to identify the tracking errors by Eq. (26).

$$\begin{cases} e_1 = I_{rd}^* - I_{rd} \\ e_2 = I_{rq}^* - I_{rq} \end{cases} \tag{26}$$

Tracking errors first derivative can be written as Eq. (27):

$$\begin{cases} \dot{e}_1 = \dot{I}_{rd}^* - \dot{I}_{rd} \\ \dot{e}_2 = \dot{I}_{rq}^* - \dot{I}_{rq} \end{cases} \tag{27}$$

The derivative of the rotor currents can be obtained referring to Eq. (24) and can be written as follows (Eq. (28)):

$$\begin{cases} \dfrac{dI_{rd}}{dt} = \dfrac{V_{rd}}{L_r} - \dfrac{R_r}{L_r} I_{rd} - \dfrac{M}{L_r}\dfrac{dI_{sd}}{dt} + \omega_2 I_{rq} + \omega_2 \dfrac{M}{L_r} I_{sq} \\ \dfrac{dI_{rq}}{dt} = \dfrac{V_{rq}}{L_r} - \dfrac{R_r}{L_r} I_{rq} - \dfrac{M}{L_r}\dfrac{dI_{sq}}{dt} - \omega_2 I_{rd} - \omega_2 \dfrac{M}{L_r} I_{sd} \end{cases} \tag{28}$$

To ensure the convergence and the stability of the system, the Lyapunov function is chosen to be a quadratic function (defined as a positive function) and is given by Eq. (29).

$$V_1 = \frac{1}{2} e_1^2 + \frac{1}{2} e_2^2 \tag{29}$$

The expression of Lyapunov derivative function is defined as negative function and it is expressed as follows (Eq. (30)).

$$\dot{V}_1 = -K_1 e_1^2 - K_2 e_2^2 \tag{30}$$

Eq. (29) can be rewritten as:

$$\dot{V}_1 = e_1 \dot{e}_1 + e_2 \dot{e}_2 \tag{31}$$

In order to guarantee a stable tracking, the Back-stepping gain coefficients K_1 and K_2 need to be positive [25].

Referring to Eqs. (30) and (31), it can be concluded that (Eq. (32)):

$$\begin{cases} e_1 \dot{e}_1 = -K_1 e_1^2 \\ e_2 \dot{e}_2 = -K_2 e_2^2 \end{cases} \tag{32}$$

Consequently, we can obtain (Eq. 33):

$$\begin{cases} -K_1 e_1 = \dot{I}^*_{rd} - \dot{I}_{rd} \\ -K_2 e_2 = \dot{I}^*_{rq} - \dot{I}_{rq} \end{cases} \tag{33}$$

Based on Eqs. (28) and (33), we can obtain (Eq. (34)):

$$\begin{cases} -K_1 e_1 = \dot{I}^*_{rd} - \frac{V_{rd}}{L_r} + \frac{R_r}{L_r} I_{rd} + \frac{M}{L_r}\frac{dI_{sd}}{dt} - \omega_2 I_{rq} - \omega_2 \frac{M}{L_r} I_{sq} \\ -K_2 e_2 = \dot{I}^*_{rq} - \frac{V_{rq}}{L_r} + \frac{R_r}{L_r} I_{rq} + \frac{M}{L_r}\frac{dI_{sq}}{dt} + \omega_2 I_{rd} + \omega_2 \frac{M}{L_r} I_{sd} \end{cases} \tag{34}$$

Finally, the rotor control voltages are given by (Eq. (35)):

$$\begin{cases} V_{rd} = L_r \left[K_1 e_1 + \dot{I}^*_{rd} + \frac{R_r}{L_r} I_{rd} + \frac{M}{L_r}\frac{dI_{sd}}{dt} - \omega_2 I_{rq} - \omega_2 \frac{M}{L_r} I_{sq} \right] \\ V_{rq} = L_r \left[K_2 e_2 + \dot{I}^*_{rq} + \frac{R_r}{L_r} I_{rq} + \frac{M}{L_r}\frac{dI_{sq}}{dt} + \omega_2 I_{rd} + \omega_2 \frac{M}{L_r} I_{sd} \right] \end{cases} \tag{35}$$

3.4 Sliding mode controller design

Thanks to its advantages such as the simplicity of the implementation, the stability and the insensitivity to external disturbances, the sliding mode controller is a widely used strategy [26]. Similar to the back-stepping controller, the aim of this strategy is to stabilize a chosen virtual control state. It rests on the stabilization of a variable error, defined as a sliding surface, by selecting the suitable control inputs. In fact, the output control parameters are obtained by the determination of two components: U^{eq} and U^{N} as given in Eq. (36).

In this section, a detailed analysis of this controller is presented in order to obtain the control output rotor voltages which regulate the output voltage and frequency of the system and maintain them constant no matter which external disturbances occur.

$$\begin{cases} V_{rd} = V^{eq}_{rd} + V^{N}_{rd} \\ V_{rq} = V^{eq}_{rq} + V^{N}_{rq} \end{cases} \tag{36}$$

In our research, the sliding surface is chosen as follows (Eq. (37)):

$$\begin{cases} S_1 = I^*_{rd} - I_{rd} \\ S_2 = I^*_{rq} - I_{rq} \end{cases} \tag{37}$$

The derivative of sliding surfaces is given by (Eq. (38)):

$$\begin{cases} \dot{S}_1 = \dot{I}^*_{rd} - \dot{I}_{rd} \\ \dot{S}_2 = \dot{I}^*_{rq} - \dot{I}_{rq} \end{cases} \tag{38}$$

Based on Eq. (28), Eq. (38) can be rewritten as follows (Eq. (39)):

$$\begin{cases} \dot{S}_1 = \dot{I}_{rd}^* - \dfrac{V_{rd}}{L_r} + \dfrac{R_r}{L_r} I_{rd} + \dfrac{M}{L_r}\dfrac{dI_{sd}}{dt} - \omega_2 I_{rq} - \omega_2 \dfrac{M}{L_r} I_{sq} \\ \dot{S}_2 = \dot{I}_{rq}^* - \dfrac{V_{rq}}{L_r} + \dfrac{R_r}{L_r} I_{rq} + \dfrac{M}{L_r}\dfrac{dI_{sq}}{dt} + \omega_2 I_{rd} + \omega_2 \dfrac{M}{L_r} I_{sd} \end{cases} \tag{39}$$

In a first step, and during the sliding mode, in order to compute the first part of the control signal, we can set forward these hypotheses (Eq. (40)):

$$\begin{cases} S = 0 \\ \dot{S} = 0 \\ V_{rd}^N = 0 \\ V_{rq}^N = 0 \end{cases} \tag{40}$$

Subsequently, equivalent voltage expressions are formulated as follows (Eq. (41)):

$$\begin{cases} V_{rd}^{eq} = L_r \left[\dot{I}_{rd}^* + \dfrac{R_r}{L_r} I_{rd} + \dfrac{M}{L_r}\dfrac{dI_{sd}}{dt} - \omega_2 I_{rq} - \omega_2 \dfrac{M}{L_r} I_{sq} \right] \\ V_{rq}^{eq} = L_r \left[\dot{I}_{rq}^* + \dfrac{R_r}{L_r} I_{rq} + \dfrac{M}{L_r}\dfrac{dI_{sq}}{dt} + \omega_2 I_{rd} + \omega_2 \dfrac{M}{L_r} I_{sd} \right] \end{cases} \tag{41}$$

In a second step, during the convergence mode, to ensure the condition $S\dot{S} < 0$, we can suppose that (Eq. (42)):

$$\begin{cases} V_{rd}^N = K_1\, sign(S_1) \\ V_{rq}^N = K_2 sign(S_2) \end{cases} \tag{42}$$

To guarantee a stable tracking, K_1 and K_2 are chosen positive constants [27].

4. Controllers management strategy

With the rapid progress of control topologies, various nonlinear control strategies such as Backstepping and sliding mode controllers whetted the interest of many researchers who attempted to develop and further enhance them. These algorithms succeeded to improve different performances of the studied system, but they remain unable to ensure optimal and safe operation of the rotor side converter. In general, the rotor side converter integrated in a DFIG based WECS is estimated at 30% of the machine nominal power which presents the main advantage of the DFIG use [12]. However, with the presence of load demand power variations and wind speed fluctuations, the rotor demanded power may exceed 30% of the DFIG nominal power. Hence, to ensure a safe functioning of the RSC and to guarantee an optimal operation mode of the DFIG, a management strategy is proposed. The wind energy can be then used effectively in order to satisfy the demand of the connected load on the one hand and to ensure security of the DFIG rotor side converter with an optimal operation on the other hand.

Based on the captured wind velocity and the load demanded power, the proposed management algorithm computes and specifies secure operation boundaries $(0.7\,\omega_{1n} \le \omega_r \le 1.3\,\omega_{1n})$.

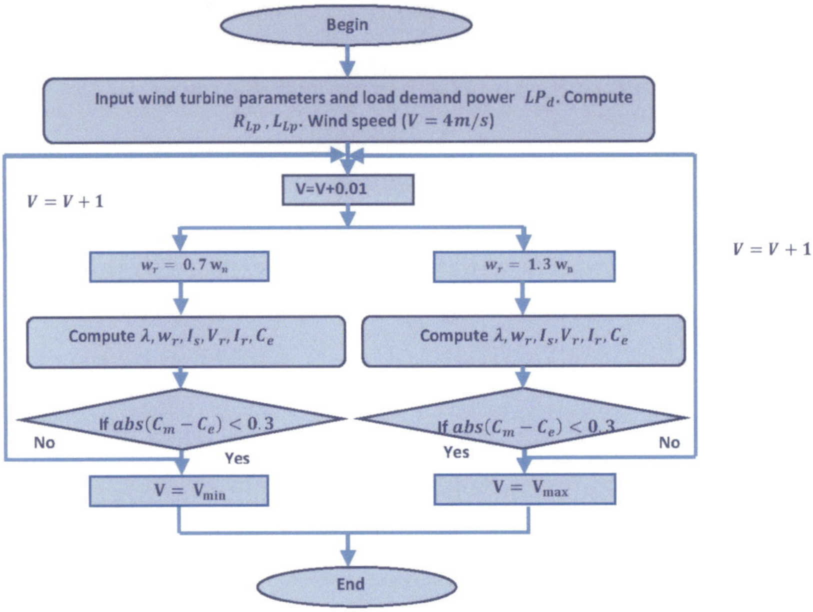

Figure 4.
Management strategy flowchart.

Thus, in order to obtain nominal stator output voltage and frequency ($V = 220V, F = 50H$), both rotor voltage and frequency change with every detected change in load demand power (Eq. (43)) to maintain a constant electromagnetic torque (Eq. (44)). As it is shown in (Eq. (43)), the rotor voltage depends mainly on the load impedance whereas, the rotor pulsation depends basically on the rotational speed of the machine.

$$\begin{cases} \overline{V}_r = \dfrac{j}{\omega_1 M} \cdot \dfrac{\overline{Z}.\overline{Z}_{sT}}{\overline{Z}_{ch}} \overline{V}_s \\ \omega_2 = \omega_1 - \omega_r \end{cases} \tag{43}$$

$$C_e = \frac{3}{2} p \frac{\overline{V}_s^2}{\omega_1} \mathfrak{Imag}\left(j \frac{\overline{Z}_{sT}}{\overline{Z}_{ch}^{\,2}} \right) \tag{44}$$

where $\overline{Z}_{LP} = R_{LP} + j\omega_1 L_{LP}$, $\overline{Z}_r = R_r + j\omega_2 L_r$, $\overline{Z}_{sT} = (R_s + R_{LP}) + j\omega_1(L_s + L_{LP})$ and $\overline{Z} = \overline{Z}_r + \frac{\omega_1 \omega_2 M^2}{\overline{Z}_{sT}}$.

The proposed algorithm can be summarized by the flowchart displayed in **Figure 4**.

The handling of the algorithm detailed in **Figure 4** allows us towards the end to obtain the speed range for every connected load. Within this framework, the main idea of the developed strategy is to detect first the load demanded power. Based on the wind turbine parameters, the connected load value and the rotor speed limits ($w_r = 0.7w_n$ and $w_r = 1.3w_n$), the wind speed limits are calculated (V_{min} and V_{max}). Otherwise, if the detected wind speed does not respect the given algorithm boundaries, the DFIG has to disconnect from the load unless the wind speed respects computed limits.

5. Simulation results and discussion

In order to analyze the system modeling, to check performances of the studied controllers and to compare the system responses using each control strategy, the proposed stand-alone wind energy conversion system based on doubly fed induction generator is implemented and tested using Matlab/Simulink environment.

The studied system rests on a wind turbine, a doubly fed induction generator and a three-phase isolated load. The DFIG parameters are obtained experimentally in the LTI, Cuffies-Soissons, France laboratory. They are exhibited in **Table 1** [28].

To satisfy the convergence conditions of the proposed system, parameters of PI controller, Backstepping and Sliding mode controllers are selected properly.

In this paper, our intrinsic purpose is to maintain constant output stator voltage and frequency under sudden variations of wind speed and load demand. Therefore, a selected profile of wind speed deduced from the proposed management strategy and load demand is applied to the system. These profiles are chosen properly in such a way that rotor side power is limited under 30% of DFIG nominal power so as to ensure the safety of the rotor side converter.

In fact, to analyze the system performances, a comparison between the three proposed strategies of control is carried out. A set of different simulation tests have been performed and carried out for 20 seconds.

The profile of wind speed profile is presented in **Figure 5**. To check the performance of the proposed model, different sudden variations are applied to the system. The wind speed varies from $7m/s$ to 15.3 m/s. For $15 \leq t \leq 20s$, the wind velocity increases to attend 15.3 m/s. At this moment, for safety reasons, the management and control strategy reacts in such a way it disconnects the DFIG from the load.

Figure 6 represents the demand of the isolated load. The demanded power of the load presents many variations. For $0 \leq t \leq 5s$, the demand of the connected load is equal to the nominal power P_n. At $t = 5s$, the demand of the load decreases and becomes $0.8P_n$. However, at $= 10s$, an increase in the demanded power from $0.7P_n$ to $1.2P_n$ is recorded. For $15 \leq t \leq 20s$, when the wind speed increases suddenly, the load is totally disconnected from the DFIG and the power transmitted to the isolated load becomes equal to zero.

To interpret properly the different results, to demonstrate the load and the wind speed variation effects on the system response and to show the response of each controller, two zooms are chosen to be presented in different figures. A zoom noted (a) is performed at $t = 5s$ to show the effect of the load demanded power variation. Moreover, at $t = 15s$ a zoom noted (b) is stated in different figures in order to

$R_s(\Omega)$	$R_r(\Omega)$	$L_s(H)$	$L_r(H)$	M(H)	P(KW)
4.9	4	0.24	0.24	0.2	1.5

Table 1.
DFIG parameters.

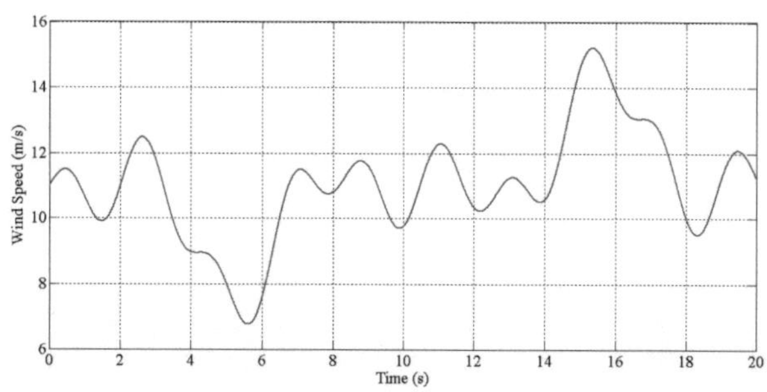

Figure 5.
Wind speed profile.

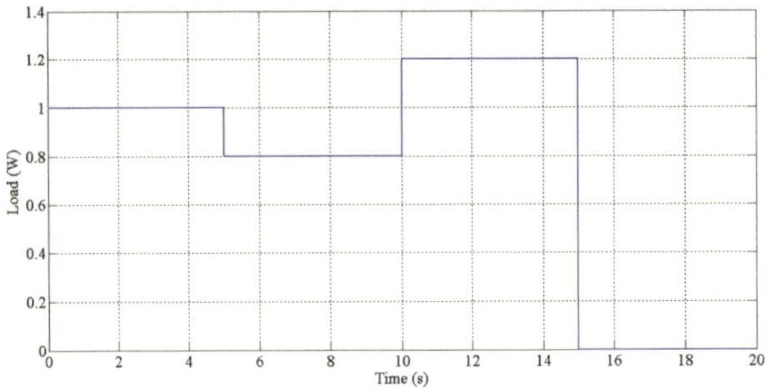

Figure 6.
Load demand variations.

demonstrate the effectiveness of the management strategy when the wind speed does not abide by computed limits.

Figure 7 plots the output stator voltages (U_{sd} and U_{sq}). We notice that the stator voltage is maintained constant in case of load demand variations (at $t = 5s$ and $t = 10\ s$) and even when a wind speed fluctuation is detected (at $t = 15s$).

Figure 8 describes the rotor voltages (U_{rd} and U_{rq}) which correspond to the control signals of the system. In each detected variation in load impedance and wind speed, controllers react by controlling rotor voltages which vary in order to maintain constant output voltage and frequency. For example, at $t = 5s$, the load demanded power changes. So, the direct rotor voltage (U_{rd}) decreases and the quadratic rotor voltage (U_{rq}) increases instantaneously to regulate the stator outputs. In fact, when the load demand varies at $t = 10s$, the control system reacts to obtain the same results

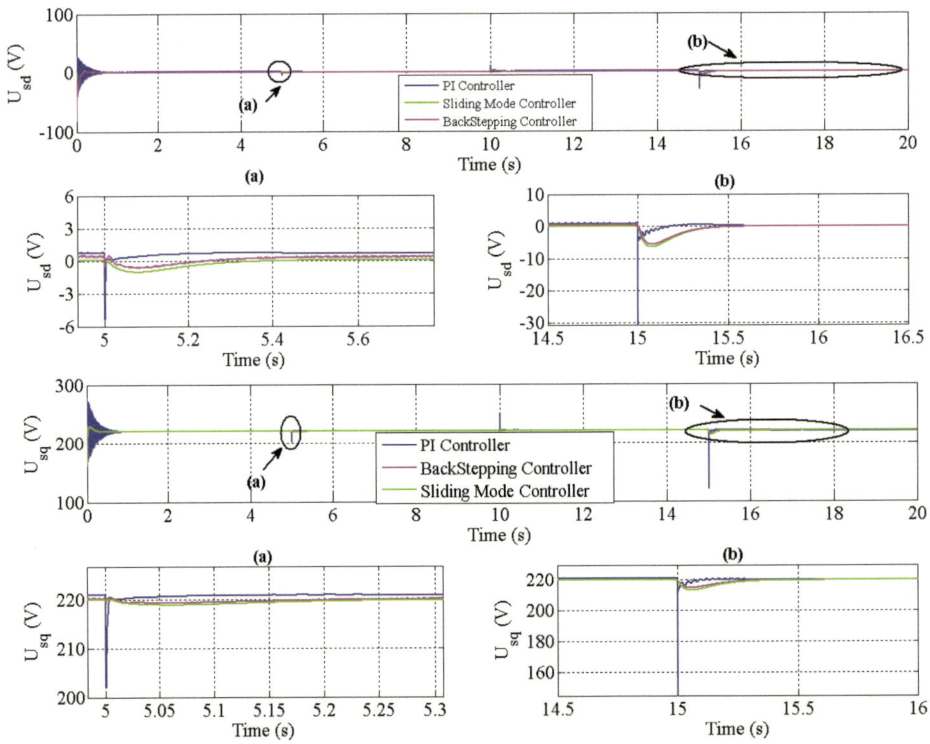

Figure 7.
The stator output voltages responses using the PI controller, the Backstepping controller and the Sliding mode controller. Two zooms are done at t = 5 s noted (a) and at t = 15 s noted (b) to see the advantages and disadvantages of each controller.

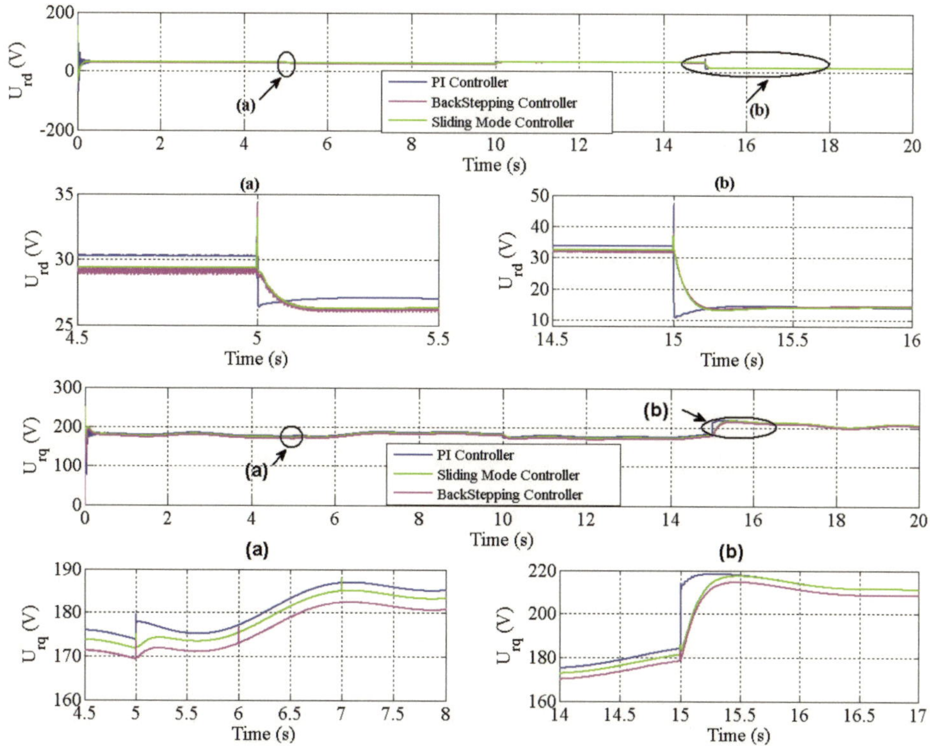

Figure 8.
The rotor output voltages responses using the PI controller, the Backstepping controller and the Sliding mode controller. Two zooms are done at t=5 s noted (a) and at t=15 s noted (b) to see the advantages and disadvantages of each controller.

in stator side. Furthermore, at $t = 0.1s$, in both **Figures** 7 and **8**, we detect the existence of a transient regime uniquely in case a PI controller is used. At this starting time, some picks are presented. However, the response of the backstepping and the sliding mode controllers are smoother and more flexible. Thus, while comparing the response of the backstepping and the sliding mode controllers, it is obviously visible in **Figures** 7 and **8** that the steady state regime is reached faster when the backstepping controller is used. In both **Figures** 7 and **8**, the presence of two zooms (a and b) permits to check the performances of the studied system, to judge the response of each controller and to select the best one. Departing from those figures (**Figures** 7(**a**) and (**b**) **and** 8(**a**) and (**b**)), in case of implementing a PI controller, we notice the presence of overshoot in stator and rotor output voltages. Thus, the robustness of the backstepping and sliding mode controllers is demonstrated.

Figure 9 highlights the evolution of the stator pulsation w_1, the rotor pulsation w_2 and the electric angular speed of the DFIG w_r. We infer that w_2 and w_r vary inversely to keep a constant stator pulsation and constant output frequency accordingly. Thus, the variation of the wind speed affects the rotor pulsation and the electric angular speed which is a normal process.

The stator output currents are displayed in **Figure 10**. It is deduced that when the power demanded by the load changes, the stator currents change in order to satisfy the demand of the load. Moreover, in zone (b), the wind speed increases and reaches 15.3 m/s. Therefore, for security reasons, the management strategy reacts to disconnect the DFIG from the load and subsequently, the power supplied to the load becomes equal to zero. At this moment, the stator currents increase and become equal to zero. However, the controller operates properly, and the stator voltage remains equal to the nominal value 220 V and remains ready for the next coupling of the DFIG and the load.

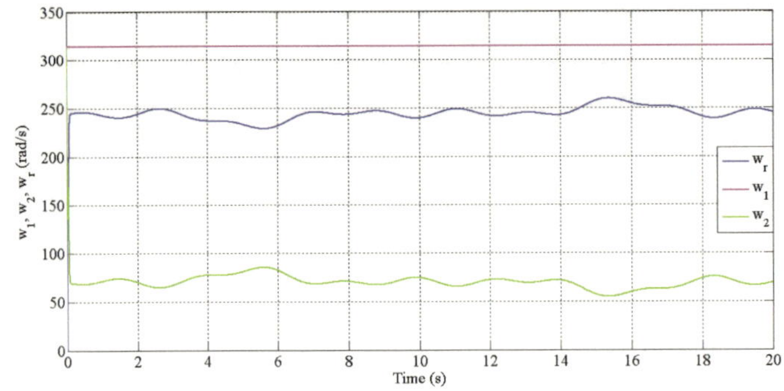

Figure 9.
Machine pulsations.

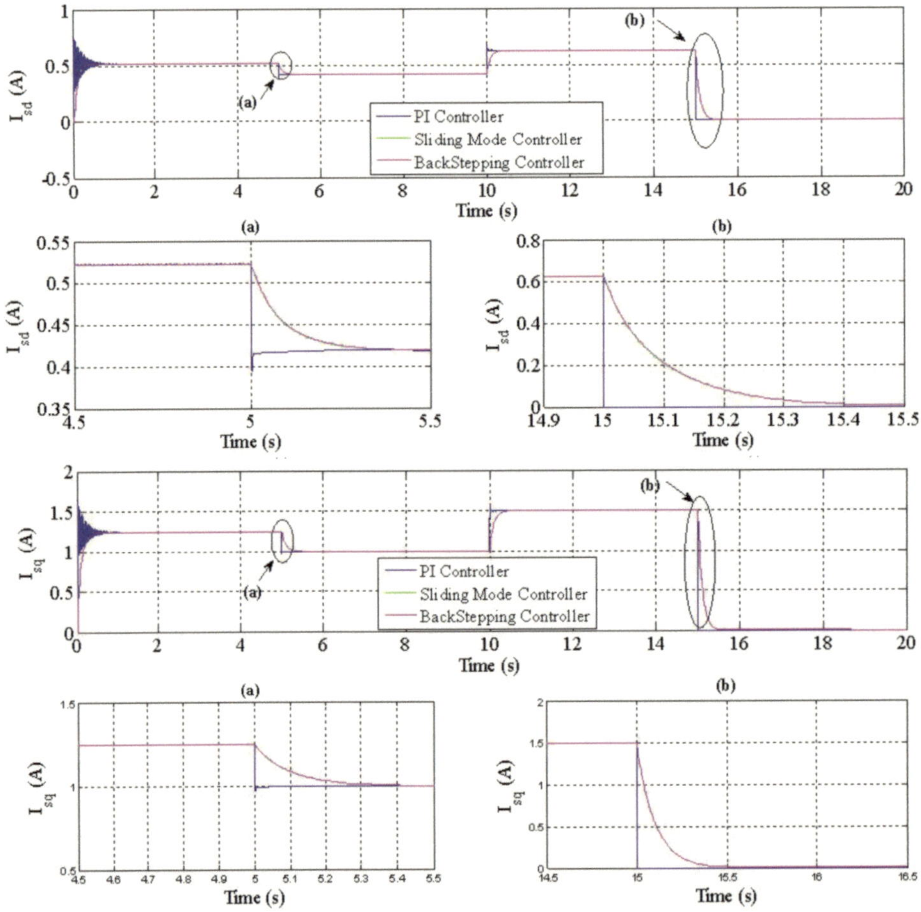

Figure 10.
The stator output currents responses using the PI controller, the Backstepping controller and the Sliding mode controller. Two zooms are done at t=5 s noted (a) and at t=15 s noted (b) to see the advantages and disadvantages of each controller.

A scrutiny of **Figures 8–10** reveals that, by comparing these three strategies, the results obtained by using the Backstepping and the sliding mode controllers are faster and more flexible than these obtained by using the PI control topology.

In order to evaluate developed control strategy performances for each load and wind speed variations, the RMSE (Root-Mean-Square Error) is calculated as follows:

Method	RMSE (V_{sd})	RMSE (V_{sq})
PI Controller	0.78	1.07
Backstepping Controller	0.45	0.32
Sliding mode Controller	0.11	0.09

Table 2.
V_{sd} and V_{sq} curves errors (Time = [0s − 5s]).

Method	RMSE (V_{sd})	RMSE (V_{sq})
PI Controller	0.64	0.87
Backstepping Controller	0.38	0.24
Sliding mode Controller	0.08	0.07

Table 3.
V_{sd} and V_{sq} curves errors (Time = [5s − 10s]).

Method	RMSE (V_{sd})	RMSE (V_{sq})
PI Controller	0.85	1.2
Backstepping Controller	0.55	0.36
Sliding mode Controller	0.13	0.11

Table 4.
V_{sd} and V_{sq} curves errors (Time = [10s − 15s]).

$$RMSE = \sqrt{\frac{1}{N}\sum\nolimits_{i=1}^{N}(x_i - \overline{x_i})^2} \tag{45}$$

Where N is the number of obtained points, $\overline{x_i}$ is the estimated value, x_i is the observed value.

Tables 2–4 sum up the RMSE of different algorithms for the same conditions (wind speed and load variations). As noticed, in each table, the RMSE of the Sliding mode controller is the smallest value. As a matter of fact, the different results obtained by this controller are close to the desired results which confirms the effectiveness of the sliding mode algorithm compared to others.

6. Conclusion

The modeling and analysis of the isolated DFIG based on WECS have been presented. The main purpose of this chapter is on the one hand to regulate the output voltage and frequency, to maintain them constant and equal to their nominal values (220 V, 50 Hz) under wind speed fluctuations and load demand variations and on the other hand to guarantee a safe operation mode of the rotor side converter through limiting the rotor side power by around 30% of the machine power. Therefore, three different system control strategies are proposed and examined. Compared to the PI controller, a stable operation of the whole system is obtained with the application of back-stepping and sliding mode controllers. However, the Sliding mode controller presents more precision and its responses are much faster than the Backstepping controller. Thus, the system performances such as precision, stability, rapidity are improved. Analysis and simulation results prove the accuracy as well as the effectiveness of both the Back-stepping and sliding mode control strategies compared to classic control strategy.

Acknowledgements

Authors of this work would like to thank University of Sfax, TUNISIA and especially the research unit CMERP for providing the facilities and research grant to achieve this research and great thanks to Prof Larbi CHRIFI-ALAOUI from LTI laboratory of University of Picardie Jules Verne and Prof Said DRID from LSPIE laboratory University of Batna for their efforts, advices and collaboration.

Nomenclature

R_s	resistance of a stator phase
L_s	stator inductance
ψ_s	stator flux
I_s	stator current
V_s	stator voltage
w_1	electric stator angular speed
w_r	electric angular speed of the DFIG
R	R is the turbine radius
V	wind speed
R_r	resistance of a rotor phase
L_r	rotor inductance
ψ_r	rotor flux
I_r	rotor current
V_r	rotor voltage
w_2	electric rotor angular speed
M	maximum coefficient of mutual induction
ρ	air density
G	gain of the gearbox

Author details

Nouha Bouchiba[1]*, Souhir Sallem[1], Mohamed Ben Ali Kammoun[1], Larbi Chrifi-Alaoui[2] and Saïd Drid[3]

1 Laboratoire Systèmes Electriques et Energies Renouvelables LSEER, National Engineering School of Sfax (ENIS), Tunisia

2 University of Picardie Jules Verne, Cuffies, France

3 LSPIE Laboratory, University of Batna, Algeria

*Address all correspondence to: bouchibanouha@gmail.com

References

[1] Ihssen Hamzaoui, Farid Bouchafaa, Abdelaziz Talha. Advanced control for wind energy conversion systems with flywheel storage dedicated to improving the quality of energy. International journal of hydrogen energy 41(2016) 20832–20846.

[2] Donghua Wang. A Novel Variable Speed Diesel Generator Using Doubly Fed Induction Generator and Its Application in Decentralised Distributed Generation Systems. Ph.D. thesis, Curtin University, School of Electrical and Computer Engineering 2012

[3] C. Evangelista, P. Puleston, F. Valenciaga. Wind turbine efficiency optimization. Comparative study of controllers based on second order sliding modes. International journal of hydrogen energy 35(2010) 5934–5939.

[4] S. Tamalouzt, N. Benyahia, T. Rekioua, D. Rekioua, R. Abdessemed. Performances analysis of WT-DFIG with PV and fuel cell hybrid power sources system associated with hydrogen storage hybrid energy system. International journal of hydrogen energy 41(2016) 21006–21021.

[5] C.A. Evangelista, F. Valenciaga, P. Puleston. Multivariable 2-sliding mode control for a wind energy system based on a double fed induction generator. International journal of hydrogen energy 37(2012)10070–10075.

[6] Mohamed Kaouane, Akkila Boukhelifa, Ahmed Cheriti. Regulated output voltage double switch Buck-Boost converter for photovoltaic energy application. International journal of hydrogen energy 41(2016)20847–20857.

[7] S. Ould Amrouche, D. Rekioua, T. Rekioua, S. Bacha. Overview of energy storage in renewable energy systems. International journal of hydrogen energy 41(2016) 20914–20927.

[8] Khouloud Bedoud, Mahieddine Ali-rachedi, Tahar Bahi, Rabah Lakel. Adaptative Fuzzy Gain Scheduling of PI Controller for control of the Wind Energy Conversion Systems. Energy Procedia 74(2015)211–225.

[9] R.Melicio, V.M.F. Mendes, J.P.S. Catalao. Comparative study of power converter topologies and control strategies for the harmonic performance of variable-speed wind turbine generator systems. Energy 36 (2011) 520–529.

[10] Rishabh Dev Shukla, Ramesh Kumar Tripathi. Isolated Wind Power Supply System using Doubl-fed Induction Generator for remote areas. Energy Conversion and Management 96 (2015) 473–489.

[11] Youcef Soufi , Sami Kahla, Mohcene Bechouat. Feedback linearization control based particle swarm optimization for maximum power point tracking of wind turbine equipped by PMSG connected to the grid. International journal of hydrogen energy 41(2016) 20950–20955.

[12] Souhir Sallem, Nouha Bouchiba, Soulayman Kammoun, Mohamed BA Kamoun. Energy management algorithm for optimum control of an off-battery autonomous DG/DFIG based WECS. The International Journal of Advanced Manufacturing Technology (2016), doi:10.1007/s00170-016-9682-1.

[13] Nouha BOUCHIBA, Souhir SALLEM, M.B.A KAMMOUN. Three-Phase Self-Excited Induction generator analysis in stand-alone mode. 6th International Renewable Energy Congress (IREC) 2015.

[14] Soro Siell_e Martin, Ahmed Chebak. Concept of educational renewable energy laboratory integrating wind,

solar and biodiesel energies. International journal of hydrogen energy 41(2016)21036–21046.

[15] Rishabh Dev Shukla, Ramesh Kumar Tripathi. A novel voltage and frequency controller for standalone DFIG based Wind Energy Conversion System. Renewable and sustainable Energy Reviews 37(2014)69–89.

[16] Gonzalo Abad, Jesus Lopez, Miguel A. Rodriguez, Luis Marrayo, Grzegorz Iwanski. Doubly Fed Induction Machine, Modeling and control for wind energy generation. IEEE Press 445 Hoes Lane, Piscatway, JJO8854.

[17] Riad Aissou, Toufik Rekioua, Djamila Rekioua, Abdelmounaı¨m Tounzi. Robust nonlinear predictive control of permanent magnet synchronous generator turbine using Dspace hardware. International journal of hydrogen energy 41(2016)21047–21056.

[18] B.Hamane, M. Benghanem, A.M. Bouzid, A. Bellabbes, M. Bouhamida, A. Draou. Control for variable speed wind turbine driving a doubly fed enduction generator using Fuzzy-PI Control. Energy Procedia 18(2012)476–485.

[19] R.K. Patnaik, P.K. Dash. Fast adaptive finite-time terminal sliding mode power control for the rotor side converter of the DFIG based wind energy conversion system. Sustainable Energy, Grids and Networks1 (2015) 63–84, 31 January 2015.

[20] F.E.V. Taveiros, L.S.Barros, F.B. Costa. Back to back converter state feedback control of DFIG (doubly fed induction generator) based wind turbines. Energy 89 (2015) 896–906,14 July 2015.

[21] Zhanfeng Song, Tingna Shi, Changliang Xia, Wei Chen. A novel adaptive control schema for dynamic performance improvement of DFIG based wind turbines. Energy 38 (2012) 104–117, 25 January 2012.

[22] Mohamed Ben Ali Kammoun. Modélisation des machines. National Engineering school of Sfax, ENIS Sfax, Tunisia.

[23] Soulaymen Kammoun. Contribution à la commande des systèmes de puissance en vue de l'intégration de l'énergie éolienne dans le réseau. Doctoral thesis 2016, National Engineering school of Sfax, ENIS Sfax, Tunisia.

[24] Badre Bossoufi, Mohammed Karim, Ahmed Lagrioui, Mohammed Taoussi, Aziz Derouich. Observer backstepping control of DFIG Generator for wind turbines variable speed: FPGA-based implementation. Renewable Energy 81 (2015) 903–97. 24 April 2015.

[25] Nihel Khemiri, Adel Khedher, Mohamed Faouei Mimouni. An adaptive nonlinear backstepping control of DFIG by wind turnines. WSEAS TRANSACTIONS on ENVIRONMENT and DEVELOPMENT. Issue 2, Volume 8, April 2012.

[26] Youcef Soufi, Sami Kahla, Mohcene Bechouat. Particle swarm optimization based sliding mode control of variable speed wind energy conversion system. International journal of hydrogen energy 41(2016) 20956–20963.

[27] V.I. Utkin, J. Güldner, and J. Shi. Sliding Mode Control in Electromechanical Systems, Florida. CRC Press, 1999.

[28] Nouha Bouchiba. Intégration du générateur à double alimentation dans un réseau isole à énergie éolienne. Doctoral thesis 2018, National Engineering school of Sfax, ENIS Sfax, Tunisia.

Chapter 10

Vertical Axis Wind Turbine Design and Installation at Chicamocha Canyon

Luis-Fernando Garcia-Rodriguez, Juan Diego Rosero Ariza, Jorge Luis Chacón Velazco and Julian Ernesto Jaramillo Ibarra

Abstract

The use of vertical axis wind turbines (VAWT) in Colombia could tackle the energy distribution difficulties as large parts of the territory are not connected to the electrical grid. The present chapter explains how to design and select an accurate VAWT for a mountain site, (the Chicamocha's canyon) by characterizing the wind energy potential, selecting the appropriate blade's airfoil, and design its corresponding blades to obtain an accurate VAWT performance. This methodology can be used to design and allocate a VAWT for residential use, as it tackles the critical point on wind energy design and selection. It is found feasible the use of wind energy at the location where the mean year density power is 485 [W/m^2], and the DU06W200 airfoil is suggested as its aerodynamic efficiency (c_l/c_d) overcomes by 14% the commonly used NACA0018. Finally, straight blades are recommended to overcome the inertial effects of the low wind velocity at the location.

Keywords: VAWT, Colombia, wind, energy, turbine

1. Introduction

There is a need of developing wind energy solutions capable to adapt fluctuating flow resources to have a diversified energy portfolio for the energy demand in Colombia [1]. The Chicamocha's canyon topography does not allow a stable electrical grid, which difficulties the incentives for tourism and commodities at the location, and the local community needs a sustainable source of energy that does not impact the environment. Therefore, this work assembles the feasibility for installing Vertical Axis Wind Turbines (VAWT) along with an optimal design.

The performance of a VAWT relies principally on its airfoil and blades, which generate lift and drag forces that take advantage of the wind kinetic energy to produce torque at the shaft of the turbine. The airfoil design and selection is an important task that depends on three main topics: wind flow conditions, airfoil shape, and modeling. The Darrieus VAWT blades design are based on lift aerodynamic forces and commonly use the commercial NACA0018 airfoil, and its performance varies according to the wind velocities [2]. Claessens [3] developed the DU06W200 airfoil for VAWT turbines, which overcomes the aerodynamic performance of the NACA0018 under high wind velocities than the calculated at Chicamocha's Canyon nature. Yarusevych and Boutilier [4] analyses a similar

 IntechOpen

Reynolds number but only one angle of attack is analyzed. Therefore, Garcia Rodriguez et al. [1] complements previous studies by increasing the range of Reynolds numbers analyzed for the DU06W200 airfoil, providing further information about the aerodynamic global coefficients and analyzing the performance of both airfoils under different attack angles.

In addition, different geometrical factors related to Power Coefficient (C_p) and VAWT turbine design is presented. These parameters vary depending on the wind characteristics of the region where the turbine is installed, therefore its analysis is fundamental to determine the VAWT performance. The 3D analysis performed allows an accurate VAWT dimensions distribution, correlates literature wind speeds (between 4 and 7 m/s) works with the Chicamocha Canyon region. Finally, different the influence of the VAWT turbine blades design is analyzed, by characterizing the Savonius rotor, Troposkien design, straight blades helical blades, observing the advantages and disadvantages of each type for the conditions of the region in question.

2. Chicamocha's canyon wind energy characterization

The Chicamocha's canyon national park, known as "PANACHI", monitor constantly the wind velocity at the canyon to control the cableway safety installed at the location. The administration of the park provided to [1] research the historical data from the year 2009 up to 2012. The wind velocity magnitude is characterized at three different locations: the two highest points of the location ("Mesa de Los Santos" and "PANACHI") and the river point.

The wind energy potential of the canyon is analyzed by using the mass conservation principle Eq. (1):

$$\frac{dm}{dt} = \rho * A * U \tag{1}$$

where ρ is the air density, U the velocity and A is the swept area. Then, the wind energy potential, P, can be expressed as kinetic energy per time unit as Eq. (2):

$$\frac{P}{A} = \frac{1}{2} * \rho * U^3 \tag{2}$$

Manwell et al. [5] establishes how significant the wind energy potential is at a selected location (**Table 1**).

The annual average wind speed and wind power density of the three locations are shown in **Table 2**. Garcia Rodriguez et al. [1] concludes that the suitable VAWT location is at Chicamocha's river due to its high wind speed, 6.9 [m/s], and the critical point is found at "Mesa de Los Santos" location.

$P/A < 100\ W/m^2$	Poor
$P/A \approx 400\ W/m^2$	Good
$P/A > 700\ W/m^2$	Excellent

Table 1.
Wind power potential criterion [5].

Place	Annual average wind speed [m/s]	Standard deviation	Annual average wind power density [W/m^2]
"Mesa de los santos"	5.9	0.736	306.188
Chicamocha's River	6.9	1.084	485.115
"PANACHI"	4.3	0.536	86.643

Table 2.
Wind power potential at Chicamocha's canyon [1].

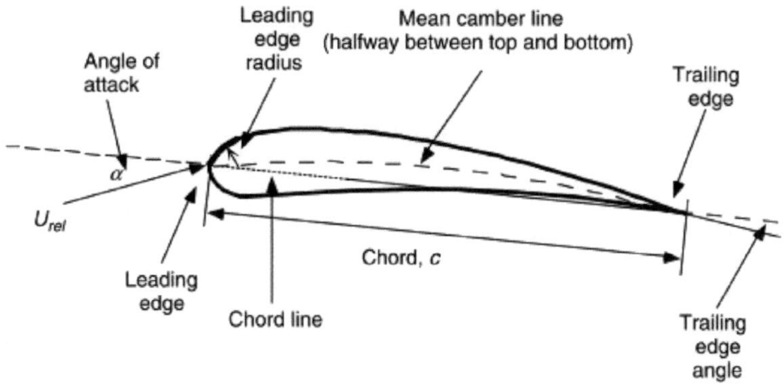

Figure 1.
Aerodynamics nomenclature [5].

3. VAWT aerodynamics

3.1 Aerodynamic airfoil and blades

An airfoil is identified using its aerodynamic parameters [5] as shown in **Figure 1**. The mean curve line is the focus midway points between the upper and lower surfaces of the airfoil. While the straight line connecting the leading and trailing edges is called the airfoil chord line, and the distance from the leading edge to the trailing edge measured along the chord line is known as the aerodynamic airfoil's chord (c). Finally, the angle of attack, α, is defined as the angle between the relative wind (U_{rel}) and the chord [6].

3.2 Lift, drag, and dimensionless parameters

The airflow over an airfoil produces a force distribution on the surface. The flow velocity increases over the convex surface resulting in lower average pressure on the "suction" side of the airfoil compared to its concave "pressure" side. Meanwhile, the viscous friction between the air and the surface of the airfoil slows the airflow to a certain point near the surface [6].

There are three forces of vital importance for aerodynamic analysis as seen in **Figure 2**, which are:

- The lifting force goes in the perpendicular direction to the incident airflow. The lift force is a consequence of the pressure differential generated between the upper and lower surfaces of the airfoil.
- Drag force is the tangential component and occurs due to friction forces on the surface of the airfoil

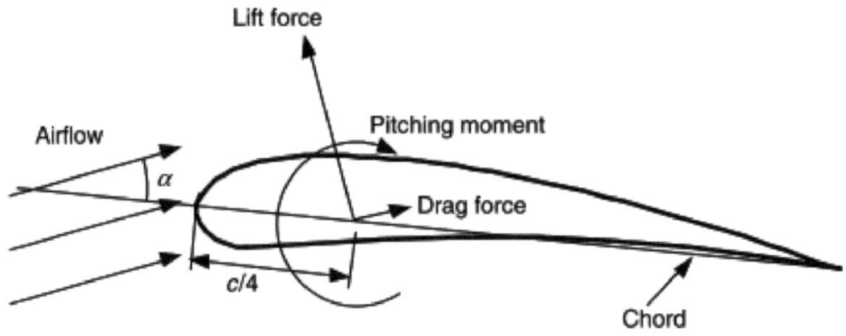

Figure 2.
Forces and moments in an aerodynamic section, an angle of attack; c, chord. The direction of positive forces and moments is indicated by the direction of the arrow [5].

- Pitch moment is defined around a perpendicular axis to the cross-section of the airfoil.

- Lift and drag coefficients: The lift and drag forces (per unit length of the blade) are usually expressed as a function of two coefficients CL and CD in Eq. (3) and Eq. (4) respectively.

$$C_L = \frac{L/l}{\frac{1}{2}\rho U^2 c} \tag{3}$$

$$C_D = \frac{D/l}{\frac{1}{2}\rho U^2 c} \tag{4}$$

where c is the chord of the blade. The lift and drag coefficients are expressed as a function of the angle of attack (γ). **Figure 3** shows the typical coefficients of wind turbine blades. Note that the CL coefficient grows approximately linearly with the angle of attack, while CD remains at a low value. For angles of attack greater than 13°, CL decreases while CD grows rapidly, and the blades go into loss.

The power output is produced through the lift force generated on the airfoil surface. As the turbine rotates, the airfoils encounter an incident wind velocity that is the vector summation of the surrounding flow velocity and the turbine rotation **Figure 4** [8].

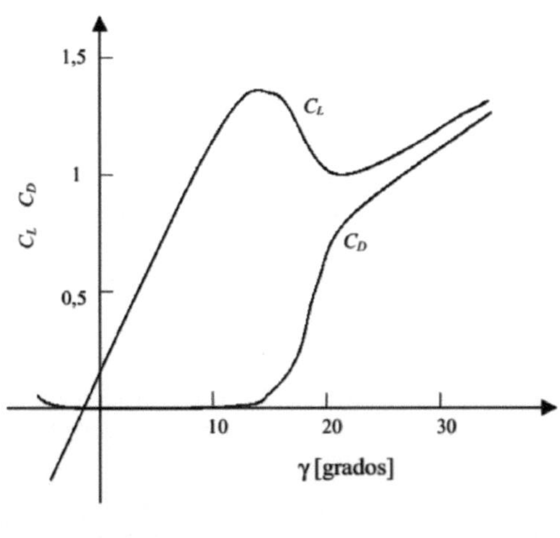

Figure 3.
Coefficients of lift and drag of a blade [7].

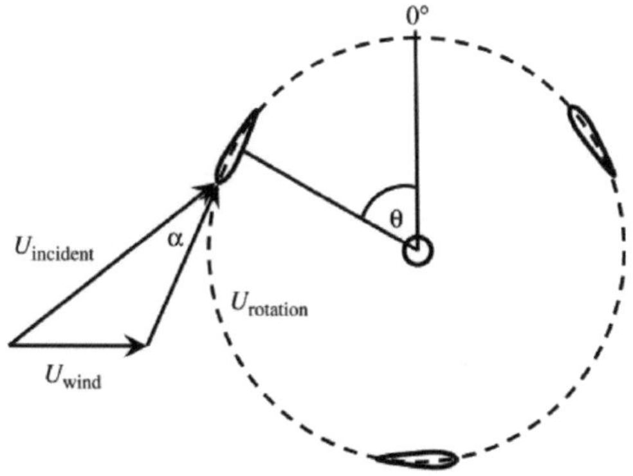

Figure 4.
Vertical axis wind turbine principle of operation. α is the relative angle of attack of the incident flow velocity U incident, and e is the angle of rotation [8].

3.3 Reynolds number (Re)

Defines the characteristics of flow conditions Eq. (5):

$$Re = \frac{UL}{v} = \frac{\rho UL}{\mu} = \frac{Inercial\ Force}{Viscous\ Force} \quad (5)$$

where μ is the fluid viscosity, U and L are the velocity and length that characterize the flow scale. These can be the inflow velocity of the flow, U_{wind} and the chord length of an airfoil. In addition to the Reynolds number of the flow conditions, the Reynolds number based on the chord c is also important.

Brusca et al. [9], defined the Reynolds number based on the chord (Re_c), and can be expressed as follow the Eq. (6)

$$Re_c = \frac{cw}{v} \quad (6)$$

where c is the chord, w is the air's relative velocity to the aerodynamic surface and v is the air's kinematic viscosity. The c can be expressed as a function of the solidity (σ) of the turbine Eq. (7):

$$c = \frac{\sigma Cp_{max}}{Nb} R \quad (7)$$

where σ is the solidity, *Cp* is the Power Coefficient, *Nb* is the number of blades and *R* is the turbine's radio. Therefore, Re_c is directly proportional to σ and *Cp* as follow Eq. (7):

$$Re_c = \frac{\sigma Cp_{max} Rw}{Nbv} \quad (8)$$

The Reynolds number strongly influences the power coefficient of a vertical-axis wind turbine. Furthermore, it changes as the main dimensions of the turbine rotor change. Increasing rotor diameter rises the Reynolds number of the blade.

3.4 Power coefficient (*Cp*)

The turbine performance is given by the power coefficient *Cp*. This coefficient represents the energy produced by the turbine as part of the total wind energy that passes through the swept area. Claessens [3], the *Cp* is represented as follow the Eq. (9):

$$C_p = \frac{P_T}{P_{wind}} = \frac{P_T}{\frac{1}{2}\rho V^3 A} \tag{9}$$

where *PT* are the total energy, ρ is the air's density, *V* is the velocity of the wind and *A* is the swept area for the turbine. Hansen et al. [10] express P_T as it shown the Eq. (10):

$$P_T = M.\omega \tag{10}$$

where *M* is the instantaneous momentum and ω is the angular velocity. Another parameter is the instantaneous moment coefficient (C_m) indicating the torque generated by the blades in the Eq. (11):

$$C_m = \frac{M}{\frac{1}{2}\rho V^2 AR} \tag{11}$$

3.5 Solidity and tip speed ratio

The solidity and Tip Speed Ratio of the turbine are directly related with the C_p as will be seen in this section, therefore, they are crucial in the design of VAWT. With the correct relation of these, it can obtain the maximum C_p. These parameters are shown below.

3.5.1 The solidity of the turbine (σ)

The solidity of the turbine (σ) it can see in Eq. (12), is defined as the developed surface area of all blades divided by the swept area [11].

$$\sigma = \frac{Nc}{R} \tag{12}$$

σ has a strong influence on VAWT performance. High solidity machines reach optimum efficiency at a low Tip Speed Ratio (λ) and efficiency drops away quickly on either side of this optimum [12].

A low solidity results in less total blade area, therefore, the blade is lighter. This benefits wind turbine performance as higher rotation speeds can be reached [6].

Paraschivoiu [11] establishes that a maximum *Cp* value rise a pick in a range of σ between 0.3 and 0.4, and then, it decreases. This peak value is not higher than *Cp* in the proposed solidity of 0.2. This statement is closely linked to the result obtained by [13], which proposes through its computational tool an optimal value of σ between 0.25 and 0.5.

In **Figure 5**, it can observe when the solidity is increased from 0.05 to 0.2, the static torque coefficient will increase by a factor of approximately 4 for an H-Darrius wind turbine. Therefore, for a high solidity, the turbine has a self-starting capability, because it has a higher static torque coefficient than the low solidity turbines [15].

Increasing σ can decrease the negative *Cp* region (i.e, when the turbine is not self-starting) and even make the *Cp* values completely positive for solidity values of 0.6 or more [3]. However, the result of this is very large blades that will increase

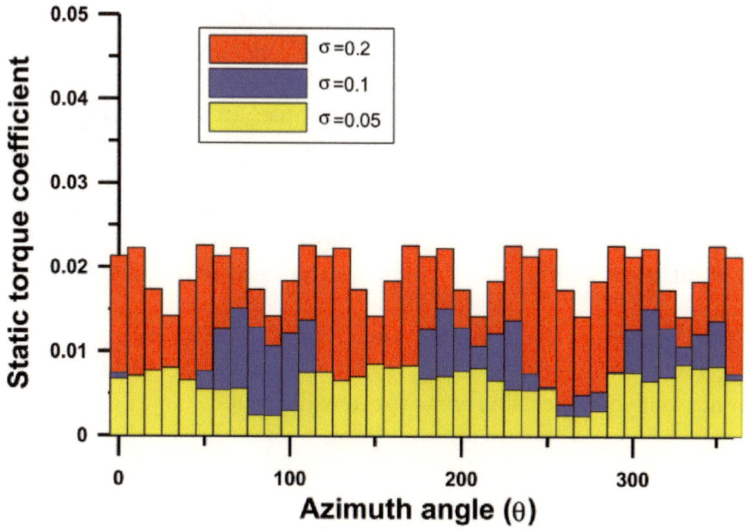

Figure 5.
Solidity effect on the static torque coefficient [14].

their manufacturing cost, so a balance must be chosen when defining the robustness of the turbine.

3.6 Tip speed ratio (λ or TSR)

The speed ratio (λ) is a ratio between the tip blade speed (ω.R) and the freestream wind velocity, and this ratio is defined as following in Eq. (13):

$$\lambda = \frac{\omega R}{V} \tag{13}$$

In **Figure 6** it can see a relation between the azimuth angle (θ), the angle of attack (α), and the speed ratio (λ), this relation is as follow in Eq. (14):

$$\alpha = \tan^{-1}\left[\frac{\sin\theta}{\lambda + \cos\theta}\right] \tag{14}$$

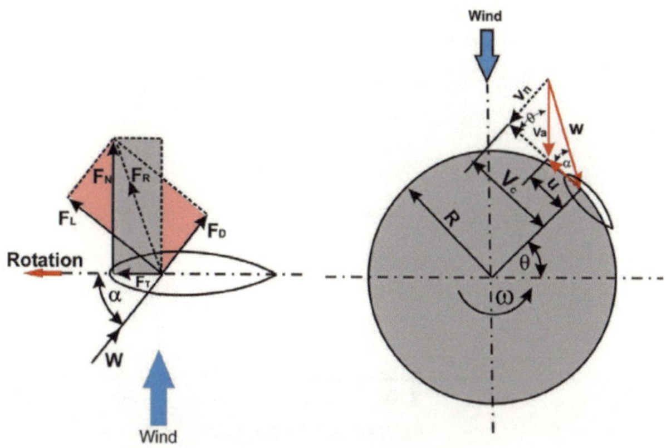

Figure 6.
Forces and velocities distribution on Darrius rotor airfoil [14].

Zouzou et al. [16] conclude in his investigation that a variable pitch VAWT has a major advantage respectively to fixed pitch VAWT in the case of high solidity rotor where the blade wake is large. That is because the pitch variation of the blade reduces flow separation and as result, the drag forces are lower. **Figure 7** shows the relationship between the drag force and the λ and the comparison between the fixed and variable pitch.

The *Cp* of a VAWT increases with an increase in the λ and reaches a peak, after which it takes a dip as larger λ are attained. **Figure 8** shows the angle of attack (α) is evaluated at different values of λ. To higher λ the value of α is smaller for a λ = 0.5 and λ = 1.5.

For VAWT λ is lower the common range is (λ = 1; λ = 5), this ranges of λ values refer to the Wind Turbine peak (Cp_{max}). **Figure 9** shows the performance of main wind turbines and some possible areas for new designs.

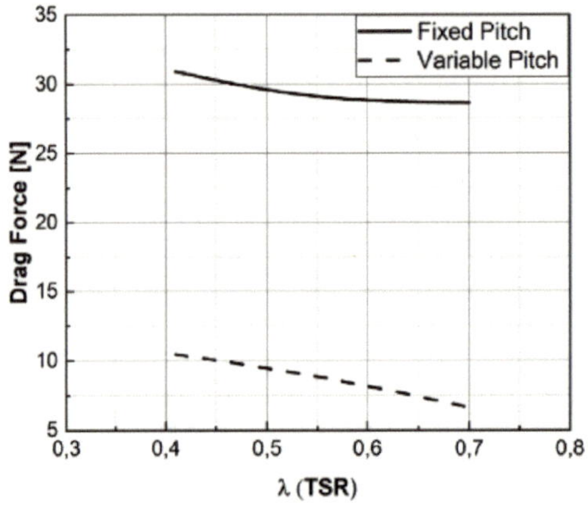

Figure 7.
The drag force of the different wind turbine configurations depends on the specific speed TSR.

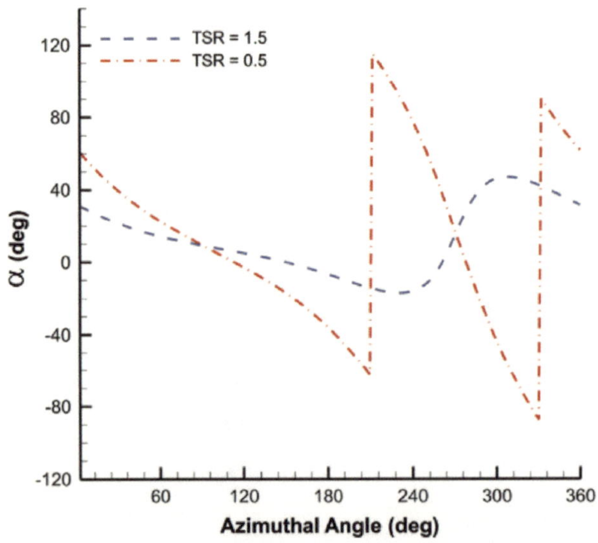

Figure 8.
Attack angle variation vs. azimuthal angle for two tip speed ratios of 0.5 and 1.5 at θp = 0° [17].

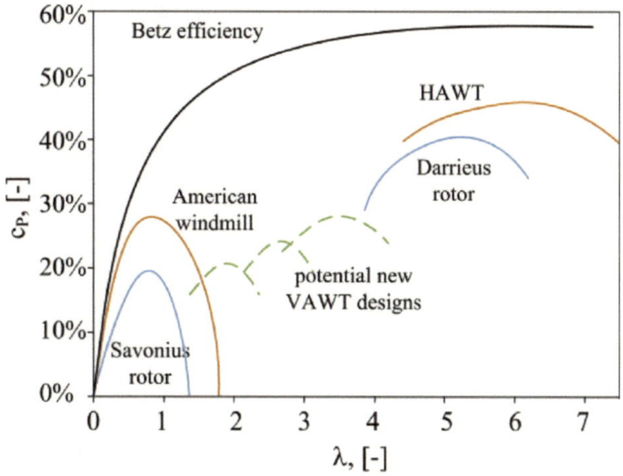

Figure 9.
Performance of main conventional wind machines and possible areas for new hybrid designs [18].

The Cp can be expressed in function of the λ and the Cm, replacing and solving Eq. (12) in (9), the Cp follow the Eq. (15) as follow:

$$Cp = \frac{Cm.R.\omega}{V} = Cm * \lambda \tag{15}$$

According to Posa [19] there is a relation between λ and the establishment of the flow downstream of a VAWT, this is related to the optimal distance between turbines in wind farm configurations. Establishing the downstream flow of a VAWT to its far-wake behavior takes a shorter distance at higher λ values.

4. VAWT design

VAWT design correlates geometrical characteristics of the rotor with the Power Coefficient (22) of the turbine. The influence of the main aerodynamic design parameters is compared with the operation of turbines [20]. This section presents the considerations and parameters necessary for the construction of VAWT turbines. The design procedure taking aerodynamics into account can be expressed as follows:

- Application and desired power
- Geometrical aspects
- Airfoil selection.

4.1 Application and desired power

The VAWT turbines have different applications [21] to generate electricity, pump water, purify and/or desalinate water by reverse osmosis, heating, and cooling using vapor compression heat pumps, mixing and aerating bodies of water; and heating water by fluid turbulence. Rathore et al. [22] suggests VAWT use on highways, in which vehicles travel at high speed in both directions producing an acceleration of the surrounding wind that can be used by turbines located in the separators.

To do so, the power (P_T) required for the application is selected. This P_T is given for a particular velocity (V), area (A), density of the air (ρ), probable Power Coefficient (C_p) and efficiencies of the mechanical components (gearbox, generator, etc.) (η) as following in the Eq. (16) [6].

$$P_T = C_p\,\eta\,\frac{1}{2}\rho V^3 A \tag{16}$$

4.2 Geometrical aspects

Among the main aspects of VAWT turbines are the chord length (c), rotor height (H), rotor diameter (D), and aerodynamic airfoil (**Figure 10**).

4.2.1 Height/diameter ratio (Φ = H/D)

The relation Φ is analyzed from the turbine shape indicating the visual proportions of the turbine. On the other hand, for a fixed sweep area, low Φ values are characteristic of turbines in which optimal flow conditions are obtained in the aerodynamics airfoil, due to large diameters that increase the peripheral speed. On the contrary, high values of Φ can be related to turbines where blade efficiency is preferred [20].

The Darrieus rotor has low aspect ratios to minimize the length of the blade and the center column for a given swept area. If the Φ is increased, then the rotor speed increases (to maintain the same relative wind speed and tip speed ratio), and torque decrease if power is constant [11].

4.2.2 Chord/diameter ratio (ξ = c/D)

High ξ values indicate that chord length is increased to improve the Reynolds number, while low values relate to rotors in which the relative wind speed increases proportionally to the relative wind speed on the aerodynamic airfoil [20].

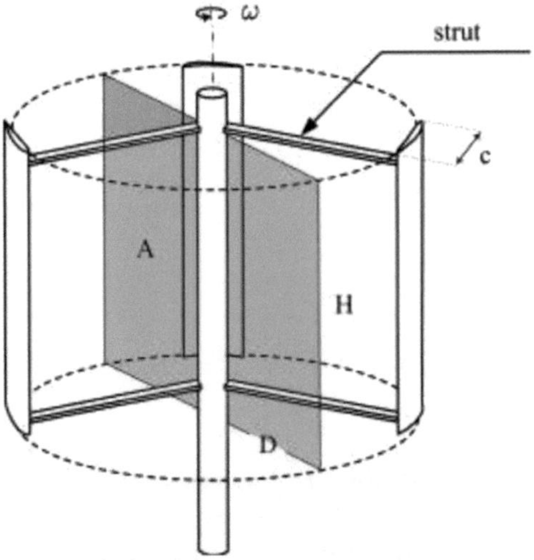

Figure 10.
Schematic view of the architecture of the Darrieus turbines [23].

4.2.3 Rotor swept area (A)

The swept area of the turbine (**Figure 10**), corresponds to the amount of air that is dragged by the turbine blades. In particular, the larger sweep areas guarantee fewer demanding limits of the turbine radius, therefore a high peripheral speed is obtained leading to a good Reynolds number on the blades.

The energy capture is proportional to the swept area and the cube of wind velocity. It is important to identify an equilibrium between energy capture and the cost of the swept area, a bigger area means more manufacturing cost of the turbine. The parameters Φ and ξ are geometric parameters that allow modifying the swept area of the turbine, they are directly related to the design of VAWT turbines.

4.2.4 Number of blades

According to Paraschivoiu [11] for given solidity, it is structurally advantageous to have fewer blades of a larger chord rather than more blades of a smaller chord. This is due to the bending stresses which are dependent on the square of the chord size whereas the aerodynamics loads are dependent on only the first power of the chord. For these reasons, the VAWT have generally two or three blades, but each design is unique for each application, therefore, it's important to analyze the relationship between the geometric parameters as the solidity and the *Cp* of the turbine. **Table 3**, can show some advantages for two and three blades in a VAWT.

4.3 Airfoil selection

The VAWT blades' performance depends largely on the airfoil behavior, which is selected or designed in terms of the wind flow conditions of the feasible location [1].

Employing CFD modeling, Garcia Rodriguez [1] found that the DU06W200 airfoil aerodynamics performance is larger than NACA0018 under the Chicamocha's canyon wind energy conditions. **Table 4** summarizes the calculated

Item	Three blades	Two blades
Construction Cost	Higher	Lower
Assembly Cost	Higher	Lower
Choice of fabrication techniques	Better	Poorer
Strength/Weight ratio	Poorer	Better
Torque ripple	Better	Poorer
Structural dynamics	Better	Poorer

Table 3.
Advantages of two or three blades [11].

Airfoil	Cl	Cd
NACA0018	0.707	0.0801
DU06W200	0.876	0.0853

Table 4.
Lift and drag coefficients of the airfoils NACA0018 and DU06W200 under Chicamocha's canyon wind speed [1].

aerodynamic coefficients of the most feasible point, proving the advantage of considering the DU06W200 airfoil [1].

5. VAWT types and selection

According to Liu et al. [24] the VAWTs are categorized as drag or lift-based devices. The first ones utilize wind drag on the blades to rotate and the last one utilizes the lift on the blades. In **Figure 11** it can observe these categories.

5.1 Drag-based turbines

The Drag-based turbines have the advantage of self-starting ability, and they are commonly found as small-sized turbines in urban and remote areas with relatively low wind speed. These turbines generally are not preferred due to high solidity, heavier weight, and low efficiency. One example of this turbine is the Savonius turbine [24], **Figure 12** shows characteristic parameters of a Savonius wind turbine with two semicircular airfoil blades.

The Savonius turbine produces high torque at low tip-speed ratios (λ) due to the large area facing the wind. The disadvantage with this turbine is that the same drag

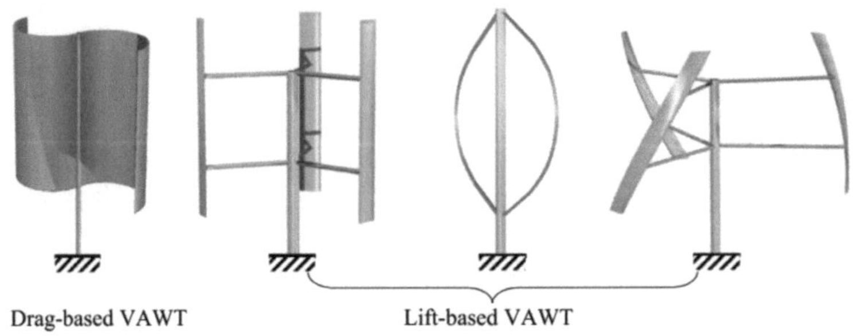

Figure 11.
Schematic view of different types of VAWTs from left to right: S-type Savonius wind turbine, straight-type, Troposkien-type, and helical-type Darrieus wind turbine [24].

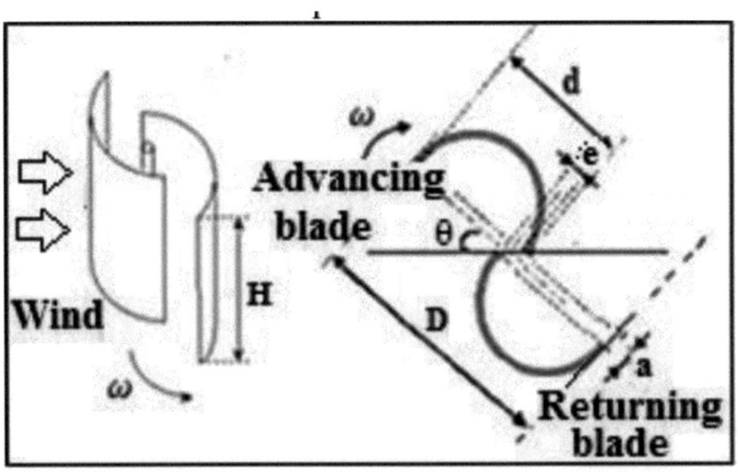

Figure 12.
Two bladed Savonius rotor [25].

of the blades which is used to produce power also works against the turbine by the returning blades, reducing the power that can be obtained [15].

According to Zemamou et al. [25] the number of blades has an important impact on the turbine performance. For obtaining the highest value of the *Cp* under the same test condition, a Savonius turbine must have two blades as shown the **Figure 13**.

5.2 Lift-based turbines

The lift-type turbine consists of airfoil sections that capture the wind energy using the lift force. This lift force produces torque on a shaft, which can then be connected to a generator to produce electricity as power output [15]. The advantage of this configuration is their simple and extruded blades, hence lower manufacturing costs [24]. The straight type, Troposkien type, and helicoidal type are examples of this configuration.

5.2.1 Straight type

These blades usually are used in small-scale, fixed pitch, rooftop designs are commercially available for domestic and other applications. The straight blades have a high value of *Cp* (0.23). This configuration can have any number of blades, from one to a configuration of five. However, the most used are two-bladed (commonly called H-type turbines) or three-bladed [26]. In **Figure 14** it can be the straight blades with two and three blades.

According to Ali and Sattar Aljabair [27] this configuration is better than the type helicoidal at low wind velocity, also, the power coefficient values for DWTs straight model with 2 blades are higher than other models as can see in **Table 5**.

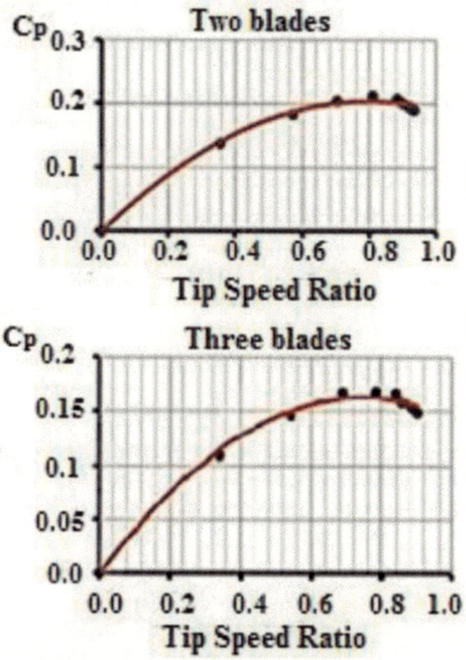

Figure 13.
The Cp variation with the TSR for two & three blades [25].

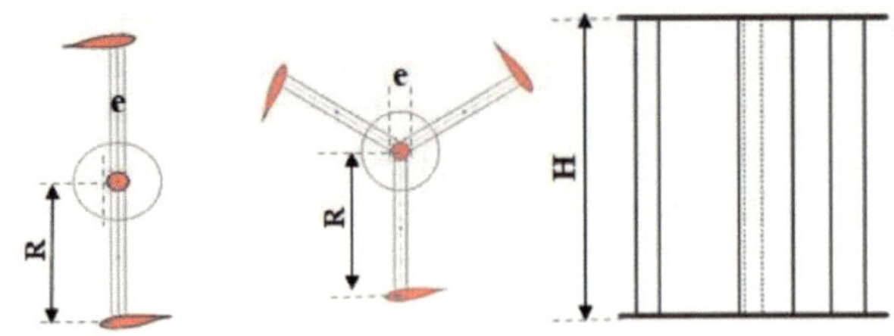

Figure 14.
Darrieus WT type straight blades with two and three blades [27].

DWT type	Number of blades	Wind velocity self-starting (m/s)						
		3	4	4.5	4.85	5.15	6.45	7.65
Straight	2	0.2495	0.2506	0.2635	0.275	0.2895	0.3076	—
	3	0.2407	0.2494	0.2606	0.2678	0.2846	0.3065	—
Twisted 70°	2	0	0	0	0	0.0372	0.0757	0.1216
	3	0	0	0	0.0195	0.0597	0.1008	0.1323
Helical 120°	2	0	0	0	0	0.0449	0.0690	0.0889
	3	0	0	0	0.0427	0.0789	0.1332	0.1465

Table 5.
The (CP) at various wind velocities for DWT models number [27].

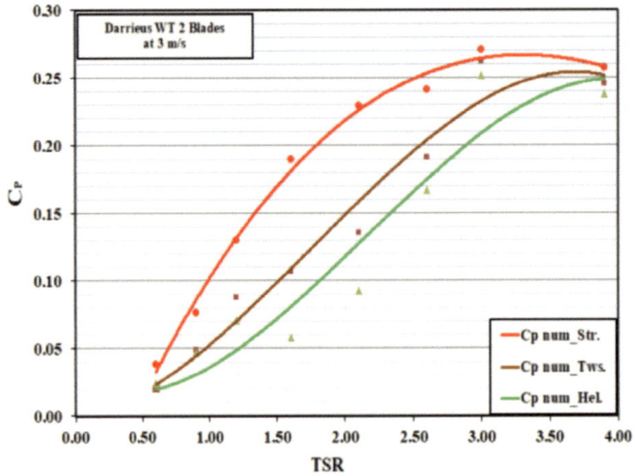

Figure 15.
The numerical relationship between CP and TSR for the DWT models has 2 blades [27].

The Straight blades present a higher value of *Cp* compared to the others as shown the **Figure 15**.

5.2.2 Troposkien

The Troposkien architecture is characterized by hub-to-hub blades, this configuration offers a lower aerodynamic drag (compared to the H-shaped one), which minimizes the bending stress in the blades [28]. In **Figure 16** it can see the Troposkien type.

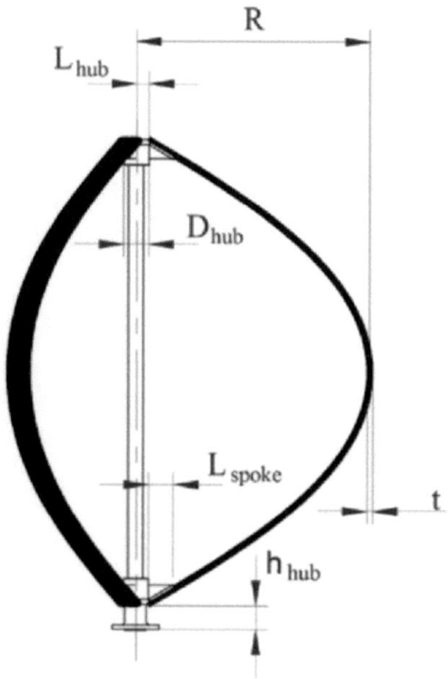

Figure 16.
The Troposkien rotor [28].

According to Battisti et al. [28] the Troposkien type is more efficient than the H-shaped configuration (two straight blades) at high values of TSR as can see in **Figure 17**. On other hand, for low values of TSR, the Troposkien present a lower *Cp* compared to the other type.

Quite similar behavior is registered for low wind velocities and a cut-in wind speed of 6–6.5 m/s is observed. For high values of wind velocity, the Troposkien is capable to generate significantly more power than the H-shaped configuration as shown in **Figure 18**. This quite different behavior could relate to the higher blade Reynolds number, which promotes an improved aerodynamic efficiency, in the investigation of [28], the radius of the Troposkien type is bigger than the H-shaped type to maintain the same rotor swept area. For this reason, the Troposkien type have a bigger Re_c and consequently bigger efficient and more power generated.

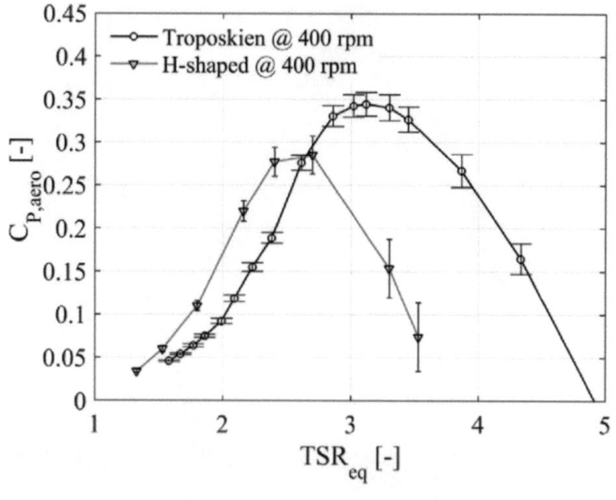

Figure 17.
Rotor power coefficients, as a function of the equatorial tip speed ratio [28].

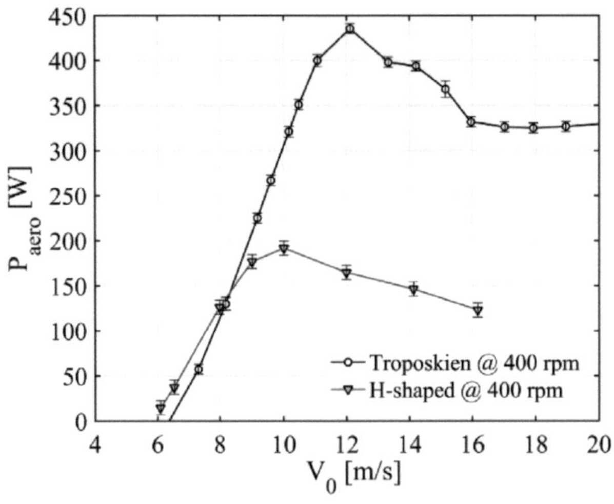

Figure 18.
Power curves for the two analyzed rotor configurations [28].

5.2.3 Helical type

Helical H-rotor distributes the blade airfoil along the rotor perimeter uniformly, thus making the swept area as well as the blade sections constant to the wind in all cases of turbine rotation [29]. In **Figure 19** it can observe the Helical type.

Tjiu et al. [29] made a comparison was made between the helical, straight, and Troposkien types. The comparison was made using 3 blades using the NACA 0015 airfoil with a TSR of 5. The behavior can be seen in **Figure 20** where it can be observed that the Troposkien typology obtained the highest fluctuation with a Cp value of approximately 0.3, the straight blades typology had a fluctuation in the Cp of 0.2 and the lowest fluctuation was obtained by the helicoidal rotor with a variation of approximately 0.03 Cp. However, despite the benefits obtained, the helical blades are more expensive to manufacture, so depending on the desired application

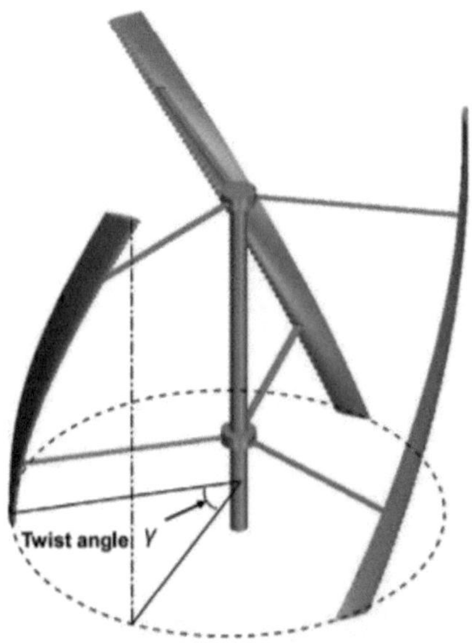

Figure 19.
Helical design [30].

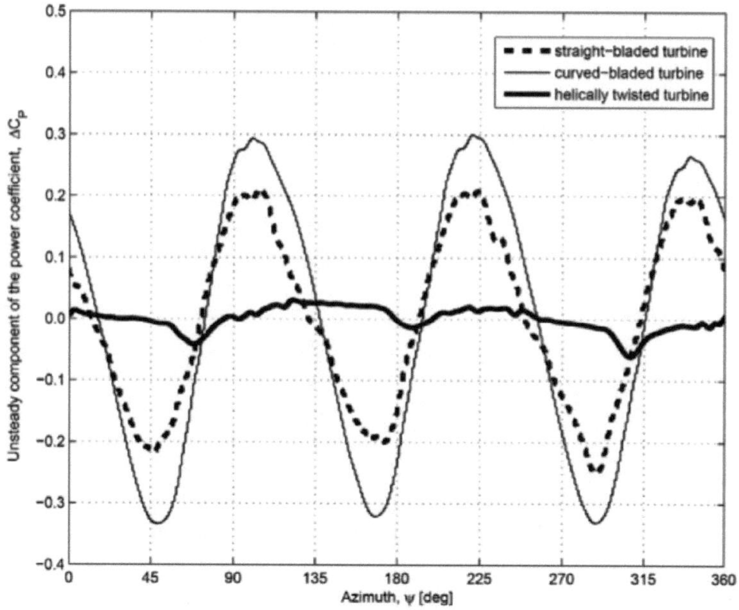

Figure 20.
Power coefficient variations of a typical Troposkien rotor, H-rotor, and helical H rotor.

and the available budget, a middle point must be chosen for the selection of the different types of rotors.

5.3 VAWT selection

A critical factor in the feasibility of power generation with VAWT turbines is the self-starting of the turbine, according to Ali and Sattar Aljabair [27], at a wind speed of 3 m/s, the VAWT with airfoil DU06W200 has the capability of self-starting as seen in **Table 6**. The straight blade type has better performance because the turbine can self-start at lower wind velocity than the others turbines.

The straight blade configuration offers the flexibility to adjust the swept area. Rotor height and diameter can be independently adjusted to suit each design. In addition, this configuration is usually mounted on a tower, which provides higher stability, lower bending, and torsional stresses on the blades compared to the Troposkien topology. Similarly, the gravity-induced bending stress is lower in the straight-bladed configuration as they are stiffer with the same chord length and thickness as the blades of a Troposkien rotor. In addition, they are vertically

DWT type	Number of blades	Wind velocity self-starting (m/s)
Straight	2	3
	3	3
Twisted 70°	2	5.75
	3	5
Helical 120°	2	6.5
	3	6

Table 6.
The wind speed at which Darrieus WT models can be DWT auto-started [27].

Wind speed	Medium	Low
Number of blades N	3	3
Rotor radius R	1 m	3 m
Rotor height H	3 m	5 m
Blade chord c	111 mm	333 mm
Rotor solidity σ	1/3	1/3
Airfoil	DU-06-W-200	DU-06-W-200
Nominal Wind speed	9 m/s	4.5 m/s
Rated power	1.5 kW	1 kW
Maximum power coefficient	0.5798	0.5996

Table 7.
Characteristics of the proposed VAWT designs [13].

positioned and suspended by supports, so they are not subjected to constant bending stress due to gravity [31].

In his investigation Meana-Fernández et al. [13] proposes an optimized design for medium and low wind speed which presents a maximum ?? of 0.5798 and 0.5996 respectively, as observed in **Table 7**.

The type of blades used by [13] were straight blades, it is observed that the proposed design presents a good performance for both medium and low wind speed. It should be noted that for low speeds, as described throughout this section, straight blades perform well without the complexity of construction and high manufacturing cost of the helical type for example, or the instability and torsional stress produced by the Troposkien type.

6. Conclusions

The current literature review analyzes a full VAWT design and installation facility by considering the site wind energy potential, the airfoil performance analysis, and the 3D blade type selection. Experimental and theoretical formulations are referenced to validate the proposed method, leading to an optimal VAWT design. It is found that Chicamocha canyon's large wind energy potential is found at its river, and the critical point is found at one of its boundaries locations ("Mesa de Los Santos"). This wind velocity is taken as a baseline point to select the airfoil and blade designs, as is the minimum value to overcome inertial effects to start VAWT rotation. Literature found that using DU06W200 airfoil, improves by 23% the aerodynamic performance of the VAWT airfoil blades, the reason why is selected to move on at the current design. Finally, the literature review shows that considering 3 straight blades on the VAWT design, complements the airfoil design and selection, as these blades have shown experimentally an accurate performance under the analyzed critical wind flow conditions. The future work will design the size of the VAWT blades and optimized the current proposes, to reach a feasible domain to be used in local facilities.

Conflict of interest

The authors declare no conflict of interest.

Author details

Luis-Fernando Garcia-Rodriguez[1*], Juan Diego Rosero Ariza[2], Jorge Luis Chacón Velazco[2] and Julian Ernesto Jaramillo Ibarra[2]

1 University of São Paulo, Brazil

2 University Industrial of Santander, Colombia

*Address all correspondence to: ingarcia1703@usp.br

References

[1] Garcia Rodriguez LF, Jaramillo JE, Chacon Velasco JL. Chicamocha canyon wind energy potential and VAWT airfoil selection through CFD modeling. Rev Fac Ing Univ Antioquia. 2019;(94):56–66.

[2] Nakano T, Fujisawa N, Oguma Y, Takagi Y, Lee S. Experimental study on flow and noise characteristics of NACA0018 airfoil. J Wind Eng Ind Aerodyn. 2007 Jul;95(7):511–531.

[3] Claessens MC. The design and testing of airfoils for application in small vertical axis wind turbines. Masters Thesis. 2006;1–137.

[4] Yarusevych S, H. Boutilier MS. Vortex Shedding of an Airfoil at Low Reynolds Numbers. AIAA J. 2011 Oct;49 (10):2221–2227.

[5] MANWELL, J.F., McGOWAN JG and RAL. Wind energy explained: theory, design and application. 2nd ed. Wiley; 2009. 677 p.

[6] Manwell J, Mcgowan J, Rogers A. WIND ENERGY EXPLAINED Theory, Design and Application Second Edition. Second. John Wiley & Sons Ltd.; 2009. 689 p.

[7] Bautista H. Control de la calidad de potencia en sistemas de conversión de energía eólica. Las turbinas eólicas. Universidad Nacional de La Plata; 2000.

[8] McLaren K, Tullis S, Ziada S. Measurement of high solidity vertical axis wind turbine aerodynamic loads under high vibration response conditions. J Fluids Struct. 2012;32:12–26.

[9] Brusca S, Lanzafame R, Messina M. Design of a vertical-axis wind turbine: how the aspect ratio affects the turbine's performance. Int J Energy Environ Eng. 2014;5(4):333–340.

[10] Hansen JT, Mahak M, Tzanakis I. Numerical modelling and optimization of vertical axis wind turbine pairs: A scale up approach. Renew Energy [Internet]. 2021;171:1371–81. Available from: https://doi.org/10.1016/j.renene.2021.03.001

[11] Paraschivoiu I. Wind Turbine Design: With Emphasis on Darrieus Concept. Schettini S, editor. Presses Inter Polytechnique. Montreal: Presses Internationales Polytechnique; 2002. 1–438 p.

[12] Edwards J. The Influence of Aerodynamic Stall on the Performance of Vertical Axis Wind Turbines [Internet]. The University of Sheffie; 2012. Available from: http://etheses.whiterose.ac.uk/2722/

[13] Meana-Fernández A, Solís-Gallego I, Fernández Oro JM, Argüelles Díaz KM, Velarde-Suárez S. Parametrical evaluation of the aerodynamic performance of vertical axis wind turbines for the proposal of optimized designs. Energy. 2018;147:504–517.

[14] Mohamed MH. Impacts of solidity and hybrid system in small wind turbines performance. Energy. 2013;57: 495–504.

[15] Mohamed MH. Impacts of solidity and hybrid system in small wind turbines performance. Energy [Internet]. 2013;57:495–504. Available from: http://dx.doi.org/10.1016/j.energy.2013.06.004

[16] Zouzou B, Dobrev I, Massouh F, Dizene R. Experimental and numerical analysis of a novel Darrieus rotor with variable pitch mechanism at low TSR. Energy [Internet]. 2019;186:115832. Available from: https://doi.org/10.1016/j.energy.2019.07.162

[17] Elkhoury M, Kiwata T, Aoun E. Experimental and numerical investigation of a three-dimensional

vertical-axis wind turbine with variable-pitch. J Wind Eng Ind Aerodyn [Internet]. 2015;139:111–123. Available from: http://dx.doi.org/10.1016/j.jweia.2015.01.004

[18] Marinić-Kragić I, Vučina D, Milas Z. Numerical workflow for 3D shape optimization and synthesis of vertical-axis wind turbines for specified operating regimes. Renew Energy. 2018; 115:113–127.

[19] Posa A. Influence of Tip Speed Ratio on wake features of a Vertical Axis Wind Turbine. J Wind Eng Ind Aerodyn [Internet]. 2020;197(April 2019): 104076. Available from: https://doi.org/10.1016/j.jweia.2019.104076

[20] Bianchini A, Ferrara G, Ferrari L. Design guidelines for H-Darrieus wind turbines: Optimization of the annual energy yield. Energy Convers Manag [Internet]. 2015;89:690–707. Available from: http://dx.doi.org/10.1016/j.enconman.2014.10.038

[21] Islam M, Ting DSK, Fartaj A. Aerodynamic models for Darrieus-type straight-bladed vertical axis wind turbines. Renew Sustain Energy Rev. 2008;12(4):1087–1109.

[22] Rathore MK, Agrawal M, Baredar P. Materials Today : Proceedings Energy production potential from the wake of moving traffic vehicles on a highway by the array of low economic VAWT. Mater Today Proc [Internet]. 2020;(xxxx). Available from: https://doi.org/10.1016/j.matpr.2020.08.638

[23] Bianchini A, Ferrara G, Ferrari L. Design guidelines for H-Darrieus wind turbines: Optimization of the annual energy yield. Energy Convers Manag. 2015;89:690–707.

[24] Liu J, Lin H, Zhang J. Review on the technical perspectives and commercial viability of vertical axis wind turbines. Ocean Eng [Internet]. 2019;182(October 2018):608–26. Available from: https://doi.org/10.1016/j.oceaneng.2019.04.086

[25] Zemamou M, Aggour M, Toumi A. Review of savonius wind turbine design and performance. Energy Procedia [Internet]. 2017;141:383–8. Available from: https://doi.org/10.1016/j.egypro.2017.11.047

[26] Aslam Bhutta MM, Hayat N, Farooq AU, Ali Z, Jamil SR, Hussain Z. Vertical axis wind turbine - A review of various configurations and design techniques. Renew Sustain Energy Rev. 2012;16(4):1926–1939.

[27] Ali, Nawfal M., Sattar Aljabair AHA. An Experimental and Numerical Investigation on Darrieus Vertical Axis Wind Turbine Types at Low Wind Speed. Int J Mech Mechatronics Eng. 2019;19(December 2019):97–110.

[28] Battisti L, Persico G, Dossena V, Paradiso B, Raciti Castelli M, Brighenti A, et al. Experimental benchmark data for H-shaped and troposkien VAWT architectures. Renew Energy. 2018;125:425–444.

[29] Tjiu W, Marnoto T, Mat S, Ruslan MH, Sopian K. Darrieus vertical axis wind turbine for power generation I: Assessment of Darrieus VAWT configurations. Renew Energy. 2015;75: 50–67.

[30] Wang Z, Wang Y, Zhuang M. Improvement of the aerodynamic performance of vertical axis wind turbines with leading-edge serrations and helical blades using CFD and Taguchi method. Energy Convers Manag [Internet]. 2018;177(May):107–21. Available from: https://doi.org/10.1016/j.enconman.2018.09.028

[31] Tjiu W, Marnoto T, Mat S, Ruslan MH, Sopian K. Darrieus vertical axis wind turbine for power generation I: Assessment of Darrieus VAWT configurations. Renew Energy. 2015;75: 50–67.